Electromechanics and Machines

MODERN ELECTRICAL STUDIES

A Series edited by
Professor G.D. Sims

*Head of Department of Electronics,
University of Southampton*

Electromechanics and Machines

R. E. STEVEN
Electrical Engineering Department
The University, Southampton

CHAPMAN AND HALL LTD.
11 NEW FETTER LANE · LONDON E.C.4.

First published 1970

Set in cold type by E.W.C. Wilkins and Associates Ltd.,
Printed in Great Britain by T & A Constable Ltd., Edinburgh.

SBN 412 09840 7

Distributed in the U.S.A.
by Barnes & Noble Inc.

Contents

Preface

This textbook is intended to provide a complete and up-to-date study of this subject for electrical, electronic and mechanical engineers. The treatment has been arranged to reflect the modern interest in machines and systems, embracing the principles of dynamic circuit theory and transient performance. But several traditional methods of analysis are retained on merit and the temptation to proceed too rapidly to the more sophisticated mathematical treatment used in generalised machine theory has been resisted.

Physical and analytical details are discussed to provide an understanding of the popular assumptions which become necessary to permit linear analyses. Some of the important topic included are; synchronous machine stability and power charts for the power system engineer, control of d.c. and induction machines using semi-conductor devices, analyses of d.c. machine types in control systems using transfer function concepts and an introduction to generalised machine theory.

An elementary knowledge of d.c. and a.c. circuit theory is assumed as well as some knowledge of electro-magnetism and differential calculus. The text will be found suitable for degree and diploma courses where this preliminary knowledge can be assumed.

It is intended as a student textbook. Various selected references are given where particular topics may be examined in wider context. Well-established notation is used and demonstrative examples with solutions are provided.

The Author wishes to acknowledge with thanks the guidance he has obtained from various papers and textbooks wherein notation and important fundamental concepts have been established. These are recommended in chapter references for further study.

‘It is as impossible to learn science from hearsay as to gain wisdom from proverbs. Ah, it is so easy to follow a line of argument and so difficult to grasp the facts that underlie it.

Our popular lecturers on physics present us with chains of deductions so highly polished that it is a luxury to let them slip from end to end through our fingers. But they leave nothing behind but a vague memory of the sensation they afforded.’

G.B. Shaw.

Units and List of Symbols

The adoption of a standard system of units and symbols has many advantages. But it does not protect the student from variations which occur in contemporary and near-contemporary literature and practice. In this textbook rationalised m.k.s. units are implied, except where specifically stated otherwise.

Attempts to provide a 'standard' notation of symbols becomes a pedantic exercise. The symbols used in this book are of common conventional usage, but the same symbol can find usage in two different contexts. The following list is provided for general guidance, in addition to local definitions within the text.

A	amplifier gain;
A	amperes (unit of current);
A	area (metres^2);
A	armature.
AT	ampere-turns
a	co-ordinate axis;
a	a phase axis;
a	real part of complex root and damping factor;
a	slot pitch;
a	parallel paths of a winding;
a	area;
a	armature.
$\overline{ac}$	specific electric loading
B	flux density (webers/m^2 = Tesla).
$\bar{B}$	specific magnetic loading.
C	capacitance (farads);
C	transformation matrix.
c	a phase axis;
c	spring compliance.
D	diameter;
D	damping coefficient.
D_e	effective damping
d	direct axis;
d	distance;
d	differential.
E	r.m.s. and d.c. voltage.

e instantaneous voltage;
e pertaining to eddy currents;
e electrical.
F force (newtons);
$F(s)$ Laplace function
$\mathfrak{F}$ magnetomotive force.
F_v viscous friction ($\propto$ velocity).
f frequency (cycles/sec = Hertz);
f instantaneous force;
f function of;
f field quantity.
G output coefficient;
G coefficient of rotational inductance;
G goodness factor.
g generator;
g damper winding on q-axis;
H magnetising force (ampere turns);
H heneries (unit of inductance);
H inertia constant.
h pertaining to hysteresis.
I r.m.s. current, d.c. current.
i instantaneous current.
J polar moment of inertia.
j mathematical operator = $(-1)^{1/2}$
K a constant
k kilo = 10^3 unit;
k coefficient of coupling;
k a constant.
L inductance (henries);
L length;
L line value.
l length
ln natural logarithmn.
M mass (kilograms);
M angular momentum;
M mutual inductance.
m phases;
m milli = 10^{-3}unit;
m metres;
m coils/phase;
m motor;

m mechanical;
max maximum value.
N magnetising turns;
N r.p.m.;
N North polarity.
n general numeric, (e.g. n^{th} harmonic);
n speed r.p.s.;
n natural (frequency).
O origin of a graph.
o subscript for no load;
o unity p.f.;
o resonant value.
P power (watts)
$\mathcal{P}$ permeance
p differential operator d/dt;
p pairs of poles;
p phase value;
p pole pitch;
Q kilovolt-amperes
q quadrature axis;
q magnetic neutral axis;
q charge (coulombs).
R ohmic resistance
$\mathcal{R}$ reluctance
r radius;
r ohmic resistance;
r rotor
S South polarity
ss steady-state
s complex frequency;
s synchronous;
s fractional slip;
s stator.
T time constant (seconds).;
T time.
Ⓣ torque (newton-metres).
t time (seconds);
t thickness;
t terminal value;
t Thevenin value;
t transpose of a matrix.

V r.m.s. or d.c. voltage (usually applied);
V unit of voltage.
v velocity (metres/sec);
v instantaneous voltage.
W energy (Joules or watt.secs.);
W power in watts.
Wb webers.
X reactance
x displacement;
x dimension;
x Steinmetz index;
x algebraic term.
Y admittance;
Y pole-pitch.
y coil-span or pitch.
Z electrical impedance;
Z number of conductors.
z_m impedance in mechanical units.

Greek:
α impedance angle = $\tan^{-1}R/X$
β a constant index
δ misalignment and load angle, (electrical degrees)
∂ partial differential
Δ elemental change
ϵ exponential, base of natural logarithmns.
ζ damping ratio
η efficiency
θ an angle; a quantity.
λ flux linkages; impedance angle = $\tan^{-1} X/R$
μ permeability; micro = 10^{-6} unit.
μ_0 magnetic space constant (= $4\pi . 10^{-7}$H/m)
ξ electric field strength (volts/metre = newtons/coulomb)
π mathematical constant.
ρ resistivity, (electrical).
σ leakage coefficient; real axis co-ordinate in complex plane.
Σ summation
ϕ instantaneous magnetic flux; phase angle.
Φ magnetic flux (webers = Tesla.m^2)
ψ mechanical angle
ω angular frequency; speed, rads/sec.

1 Introduction

Energy derived from natural resources is commonly utilised at points remote from the source and in forms which usually differ from the original. Conversion to other forms of energy and the efficient transmission of the energy over long distances, sometimes to relatively inaccessible points of application, is possible using electrical energy as the intermediary.

Energy is converted to electrical energy by 'generators' of various kinds. The electricity is conveyed by circuits which may include resistance, inductance, capacitance, transformers, amplifiers or rectifiers.

Utilization occurs using electromechanical devices such as motors, relays, loudspeakers, indicating meters etc., which convert the electrical energy into the desired form. Other appliances which utilise electrical energy as heat, light or in chemical processes are not considered here.

A device which converts energy from one form to another is called a 'transducer'. Electrical transducers are essentially part of a system. They may be classified as;
(*i*) those which provide a steady transfer of energy and operate normally in steady-state at high efficiency,
(*ii*) those which transfer information and may operate intermittently, using only small amounts of energy.

A few principles determine the behaviour of most of these devices. Some aspects of these fundamental ideas are reviewed. Various forms of study and analysis have been developed, each possibly with an emphasis appropriate to a particular field of application, but not exclusively so, e.g. design, analysis of systems, utilization, etc.

The fundamental force law

This is commonly referred to as the '*Bli*' relation. It is derived from experimentally observed facts.

The force between two long parallel conductors, d m apart and carrying currents I and I' respectively is:

$$f = \frac{\mu_0 I I'}{2\pi d} \text{ newtons/metre of parallel run,}$$

where $\mu_0 = 4\pi \cdot 10^{-7}$ H/m in m.k.s. rationalised units.

This situation can be interpreted as a current-carrying conductor, of length l metres, located in the magnetic field of density B Wb/m² due to the other conductor. Now since

$$B = \mu_0 H = \mu_0 \frac{I'}{2\pi d}$$

at a radial distance d m from a conductor carrying I' amps.

Hence $$f = BlI \text{ newtons.}$$

More generally we note that l and B are vectors. The cross product $\mathbf{l} \times \mathbf{B}$ gives a resultant vector mutually at right angles and the 'right-hand rule' for vectors indicates the correct directions. The magnitude of the product vector is $|F| = IlB \sin\alpha$ where α = angle from l to B.

In most transducers the conductor is at right angles to the field and the simple relation holds; but in general,

$$\mathbf{F} = I\,\mathbf{l}\,\mathbf{B} \text{ newtons.}$$

This is the basic law underlying the 'motor' principle. N.B. A less elegant but useful alternative memory aid for determination of the relative directions of Current, Field and force, (or Motion), is Fleming's Left Hand Rule for motors, viz. the seCond finger, Forefinger, and thuMb held mutually at right angles indicate the corresponding quantities.

The force on a charge moving across a magnetic field follows from the above equation. Since current = coulombs/sec we can write:

$$\mathbf{f} = q\,\mathbf{v}\,\mathbf{B} \text{ newtons,}$$

where q = charge in coulombs and v = velocity in m/sec. This equation is of interest in cathode-ray and plasma jet (M.H.D.) applications.

The alignment principle

It is important to note that force and torque are static effects. Energy conversion does not occur until the point of application of the force moves and work is done.

In electrical machines two systems of current-carrying conductors are arranged to provide relative motion and energy conversion arising from mutual forces due to the '*Bli*' phenomenon. But the force and motion must be maintained to provide continuous energy conversion. The principle whereby magnetic axes tend to align with each other is examined to see how this is achieved.

Consider two coils having magnetic fields of m.m.f.'s $\mathcal{F}_1$ and $\mathcal{F}_2$, (Fig. 1.1). Motion will tend to occur to remove any misalignment, δ, and the restoring component of torque $\propto \mathcal{F}_1 \mathcal{F}_2 \sin\delta$. This is maximum when $\delta = 90°$. For continuous motion a misalignment must be maintained. In the d.c. motor this is achieved by supplying the moving armature winding from fixed brushes on a commutator. The arrangement maintains the armature-field axis at 90° to the stationary main-field axis, also providing maximum torque conditions, (Fig. 1.2). It is not necessary that one of the field axes should be stationary, only that a relative misalignment between the axes be maintained. How this is obtained in a.c. machines is explained later.

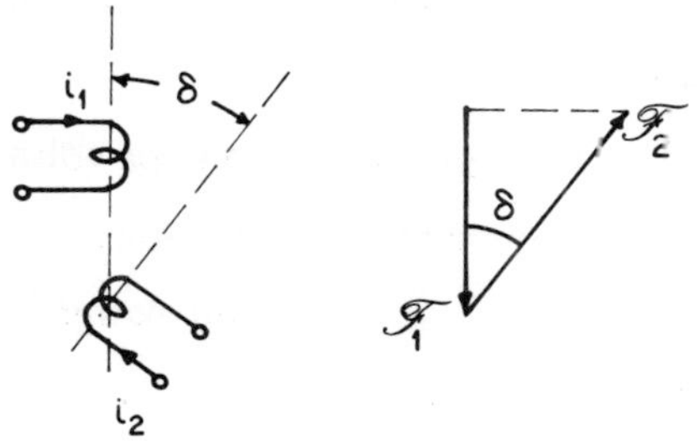

Fig. 1.1.

Electromagnetic induction

Electromagnetic induction may be viewed as a 'flux-cutting'

the flux due to relative physical movement of the coil and the field-producing magnet. A change in the flux threading the coil can be achieved by either process.

If we define Self-inductance, L = flux linkages/amp

$$\text{then } e = -L \frac{di}{dt}.$$

Dynamic response

The supply of energy to a device will result in a response dictated by the parameters of the device. Energy storage elements may be present, for example masses and springs in mechanical systems and inductors and capacitors in electric systems. Then a 'transient' period occurs before the final 'steady-state' condition. The response is the solution of a differential equation, for example;

Consider a voltage E applied to a series circuit with R and L and causing an instantaneous current i. Using Kirchhoff's Law,

$$L \frac{di}{dt} + iR = E.$$

A similar form of equation is obtained for a simple mechanical system having mass M and viscous friction F_v, and the mass has an instantaneous velocity v due to applied force F. Summing the components of force,

$$M \frac{dv}{dt} + F_v v = F.$$

The response in either case consists of the sum of the natural or transient response and the steady-state response. For the electrical case the natural response is obtained from,

$$L \frac{di}{dt} + iR = 0$$

that is, by making the forcing function, E = zero. The solution requires that the function i and its derivative be of the same form and an exponential function satisfies this condition. Hence we suggest a solution, $i = A\epsilon^{pt}$. Substitution in the natural response equation gives,

$$LAp\epsilon^{pt} + RA\epsilon^{pt} = 0$$

$$Lp + R = 0$$

$$\therefore \quad p = -\frac{R}{L}$$

$$\text{and} \quad i_n = A\epsilon^{-Rt/L}.$$

For the forcing function E, the final steady-state current,

$$I = \frac{E}{R}. \text{ Hence } i = I + i_n$$

$$= \frac{E}{R} + A\epsilon^{-Rt/L}.$$

The constant A is obtained by substituting initial conditions, thus if at $t = 0$, $i = 0$ then,

$$0 = \frac{E}{R} + A$$

$$A = -\frac{E}{R}$$

$$\therefore \quad i = \frac{E}{R}(1 - \epsilon^{-Rt/L})$$

Now $$L\frac{di}{dt} + iR = E = IR$$

$$L\frac{di}{dt} = R(I - i)$$

$$\therefore \quad \frac{di}{dt} = \frac{R}{L}(I - i)$$

$$= \frac{(I - i)}{T} \text{ where } T = \frac{L}{R}.$$

T is termed the *Time Constant* since if, at any instant, the current continued increasing at the instantaneous rate, the time to reach the final steady-state value would be T sec. Note also that when $t = T$,

$$i = I(1 - \epsilon^{-1})$$

$$= I(1 - 0{\cdot}368)$$

$$= 0{\cdot}632I.$$

Hence the time constant is also the time for the current to reach 63·2% of its final value when it is increasing exponentially.

It follows that for any exponential function the greater the time constant the longer it takes for the function to reach its final value. (Fig. 1.3).

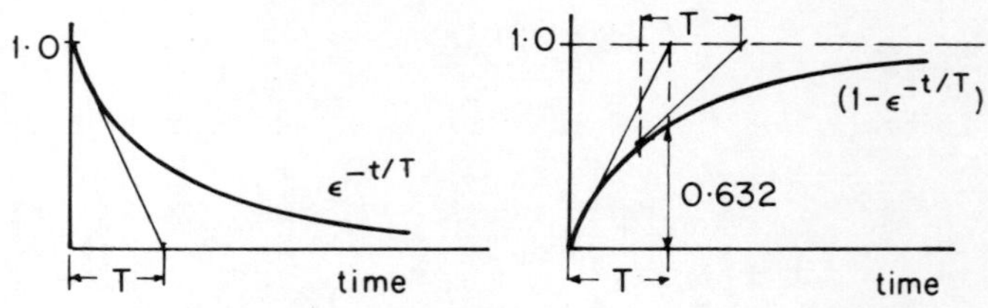

Fig. 1.3.

In the above example only one energy storage element, namely Inductance, is present and the response is obtained by a first order differential equation. The form of solution has general application in the mathematical sense. The solution in this case may be expressed:

$$i = I\,(1 - \epsilon^{-t/T}.)$$

The response of a system or device having two types of energy storage element is given by the solution of a second-order differential equation. Thus if a voltage is applied to a circuit having resistance, inductance and capacitance in series, the response equation can be written:

$$L\frac{\mathrm{d}i}{\mathrm{d}t} + iR + \frac{1}{C}\int i\cdot\mathrm{d}t = E$$

$$\text{then}\quad L\frac{\mathrm{d}^2 i}{\mathrm{d}t^2} + R\frac{\mathrm{d}i}{\mathrm{d}t} + \frac{1}{C}\cdot i = 0$$

dividing across by L,

$$\frac{\mathrm{d}^2 i}{\mathrm{d}t^2} + \frac{R}{L}\frac{\mathrm{d}i}{\mathrm{d}t} + \frac{1}{LC}\,i = 0.$$

Assuming a solution of the form, $i = A\epsilon^{pt}$ and substituting gives,

$$p^2 + \frac{R}{L}p + \frac{1}{LC} = 0.$$

This equation, sometimes termed the 'characteristic equation', is a quadratic with two solutions, say p_1 and p_2, giving

$$i_n = A_1\epsilon^{p_1 t} + A_2\epsilon^{p_2 t}$$

where $$p_1 = -\frac{R}{2L} + \sqrt{\left(\frac{R^2}{4L^2} - \frac{1}{LC}\right)}$$

and $$p_2 = -\frac{R}{2L} - \sqrt{\left(\frac{R^2}{4L^2} - \frac{1}{LC}\right)}.$$

Hence if $\frac{R^2}{4L^2} - \frac{1}{LC} > 0$ the roots are real and unequal.

$\frac{R^2}{4L^2} - \frac{1}{LC} = 0$ the roots are real and equal,

$\frac{R^2}{4L^2} - \frac{1}{LC} < 0$ the roots are complex.

Now let $p_1 = -a + j\omega$ where $a = \frac{R}{2L} = \frac{1}{T}$

$p_2 = -a - j\omega$ and $\omega = \sqrt{\left(\frac{1}{LC} - \frac{R^2}{4L^2}\right)}$.

Then since Euler's equation gives $\epsilon^{j\omega t} = (\cos\omega t + j\sin\omega t)$

$$\begin{aligned} i_n &= A_1 \epsilon^{(-a+j\omega)t} + A_2 \epsilon^{(-a-j\omega)t} \\ &= \epsilon^{-at}[(A_1 + A_2)\cos\omega t + j(A_1 - A_2)\sin\omega t] \\ &= \epsilon^{-at}\sqrt{(B_1^2 + B_2^2)}\sin(\omega t + \theta) \end{aligned}$$

where $B_1 = (A_1 + A_2)$ and $B_2 = (A_1 - A_2)$; $\theta = \tan^{-1} B_1/B_2$.

Hence the nature of the response will be dictated by the relative magnitudes of R, L and C. When the roots are real the response will be exponential in character. When the roots are complex, the response will be an exponential form superimposed on an oscillatory response, Fig. 1.4(a).

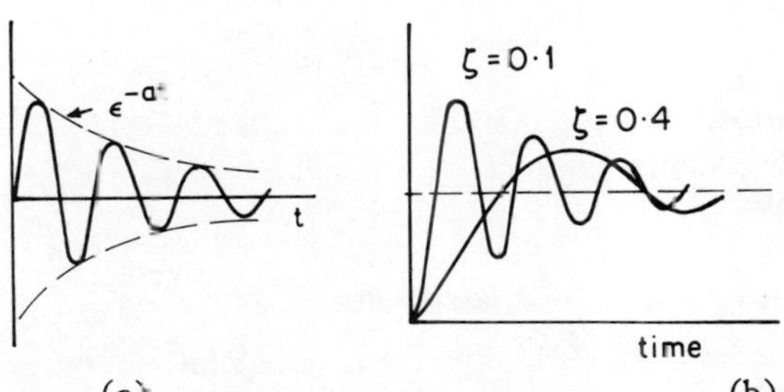

(a) Fig. 1.4. (b)

The measure of exponential decay is determined by $a = R/2L$.

If R is zero, a continuous oscillation, or undamped condition occurs and the natural frequency of oscillation,

$$\omega_n = \sqrt{\frac{1}{LC}}.$$

ω decays in value as R increases, reaching a critical value when,

$$R^2/4L^2 = 1/LC$$

$$\text{or, } R_{\text{critical}} = 2(L/C)^{\frac{1}{2}}$$

at which value no oscillation occurs. In other words, R acts as a damping agent. A measure of damping is expressed,

$$\text{Damping Ratio, } \zeta = \frac{\text{Actual damping}}{\text{Critical damping}} = \frac{R}{R_{\text{crit}}} = \frac{R}{2(L/C)^{\frac{1}{2}}} = \frac{a}{\omega_n}$$

Then, Damped Angular Frequency,

$$\omega_d = (1/LC - R^2/4L^2)^{\frac{1}{2}}$$
$$= \omega_n(1 - \zeta^2)^{\frac{1}{2}} = (\omega_n^2 - a^2)^{\frac{1}{2}}.$$

When $\zeta < 1$ the system is underdamped,

$= 1$ the system is critically damped,

> 1 the system is overdamped.

These concepts have general application to 2nd order response equations and the characteristic equation can be expressed in general form,

$$p^2 + 2\zeta\omega_n p + \omega_n^2 = 0.$$

As before, the complete solution is obtained by adding the steady-state and natural solutions, Fig. 1.4 (b).

$$i = I_{ss} + i_n$$

and the values of the constants A_1, A_2 etc. are obtained by substituting known terminal conditions, e.g. at $t = 0$ and $\mathrm{d}i/\mathrm{d}t = 0$.

Dynamics of elementary transducers

The principles reviewed may now be combined to explain the performance of certain elementary types of transducers.

(a) The simple linear motor.

Consider a straight conductor of mass M, having an effective length l and ohmic resistance R when placed at right angles on parallel conducting rails. The conductor is free to move along the rails against viscous friction F_v ($\propto$ velocity). A uniform magnetic field of flux density B is arranged at right angles both to the rails and to the conductor. When a voltage E is applied to the rails, an instantaneous current i flows and a force Bli acts on the conductor. At time $t = 0$, $i = E/R$ amps.

The conductor accelerates to instantaneous velocity v and a voltage is induced in the conductor, opposing the applied voltage E, and given by,

$$e_g = Blv \text{ volts.}$$

Then the instantaneous current

$$i = \frac{E - e_g}{R} \text{ amps.}$$

Equating instantaneous forces on the conductor;

$$M\frac{dv}{dt} + F_v \cdot v = Bli$$

$$= Bl\left(\frac{E - e_g}{R}\right) = Bl\left(\frac{E - Blv}{R}\right)$$

$$= \frac{EBl - B^2l^2v}{R}.$$

Re-arranging the equation;

$$M\frac{dv}{dt} + \left(F_v + \frac{B^2l^2}{R}\right)v = \frac{EBl}{R}.$$

The steady-state solution occurs when $\frac{dv}{dt} = 0$,

$$\therefore v_{ss} = \frac{EBl}{R(F_v + B^2l^2/R)}.$$

Hence the solution for instantaneous velocity is:

$$\underline{v = v_{ss}(1 - \epsilon^{-t/T})}$$

where

$$T = \frac{M}{(F_v + B^2l^2/R)}.$$

(*b*) *The rotational motor.*

Because of limitations in continuous or steady-state operation of the linear motor, the rotational motor is generally used for the steady transfer of electrical energy to mechanical energy. Study of the dynamics of the machine will be seen to be virtually identical to those of the linear machine, except that velocity is replaced by rotational velocity and force is replaced by torque.

Conductors carrying the supply current are housed in slots on the surface of a cylindrical air gap. [See Figs. 4.1 and 4.2 (a)]. The current is led to conductors on the 'rotor' via sliding contacts, termed 'brushes' operating on slip-rings, or in the d.c. case, on a commutator. The magnetic field is provided by electromagnets or 'poles' supported by the frame or 'stator' of the machine. Variations of these physical features will be considered in detail later.

Using the fundamental force law, and assuming d.c., inst. torque on the rotor due to current i in one conductor,

$$= Bli \cdot r \text{ newton metres}$$

If Z = number of effective conductors,

$$\text{Ⓣ}_e = Bli \cdot rZ$$

$$= KBi \text{ where } K = lrZ.$$

Equating inst. mechanical and electrical forces,

$$J\frac{d\omega}{dt} + F_v \cdot \omega = KBi$$

As before, when the conductors acquire an instantaneous angular velocity, ω, a voltage is induced according to the Blv law, opposing the applied voltage, E. The induced voltage for Z effective conductors in series is:

$$e_g = Blr\omega Z$$

$$= KB\omega$$

Hence $$i = \frac{E - e_g}{R} = \frac{E - KB\omega}{R}$$

where R = resistance of conductors in series.

Substituting in the dynamic equation:

$$J\frac{d\omega}{dt} + F_v \cdot \omega = KB\left(\frac{E - KB\omega}{R}\right).$$

Re-arranging,

$$J\frac{d\omega}{dt} + \left(F_v + \frac{K^2B^2}{R}\right)\omega = \frac{KBE}{R}.$$

The steady-state solution occurs when $d\omega/dt = 0$, viz;

$$\omega_{ss} = \frac{KBE}{R(F_v + K^2B^2/R)} \text{ radians/sec}$$

and the solution for the instantaneous velocity is:

$$\underline{\omega = \omega_{ss}(1 - \epsilon^{-t/T})}$$

where $T = JR/(F_v R + K^2B^2)$.

Note: (*i*) Steady-state speed is proportional to supply voltage, E.
(*ii*) Since F_v will be small, steady-state speed is approximately inversly proportional to field, B.
(*iii*) At starting, before e_g has built up, it is necessary to limit the current by a 'starting resistance', in series with the armature.

(*c*) *The positional device.*

A different class of device occurs if the motion resulting from the force or torque developed is limited by a restraining force. Loudspeakers and indicating meters are examples, wherein the force developed by a current-carrying coil, located in a uniform magnetic field, is counter-balanced by a spring constraint. The moving-coil type meter, or D'Arsonval movement, is an example.

The instantaneous torque, acting on the moving element, due to instantaneous current i in the coil can be written:

$$\textcircled{T} = KBi$$

where K is a constant determined by physical features of the moving element. Torque arising from angular displacement θ against spring constraint

$$= K_s\theta$$

where K_s = spring constant.

Instantaneous torques acting on the moving element;

$$J\frac{d^2\theta}{dt^2} + F_v \cdot \frac{d\theta}{dt} + K_s\theta = KBi$$

where J = moment of inertia of moving element, F_v = viscous friction.

If a constant voltage E is applied to the coil;

$$i = \frac{E - e_g}{R} \quad \text{and} \quad e_g = KB\frac{d\theta}{dt}$$

where R = ohmic resistance of coil. Then,

$$J\frac{d^2\theta}{dt^2} + F_v \cdot \frac{d\theta}{dt} + K_s\theta = KB\left(\frac{E - e_g}{R}\right)$$

$$J\frac{d^2\theta}{dt^2} + \left(F_v + \frac{K^2B^2}{R}\right)\frac{d\theta}{dt} + K_s\theta = \frac{KBE}{R}.$$

If the coil is supplied from a constant-current source;

$$J\frac{d^2\theta}{dt^2} + F_v \cdot \frac{d\theta}{dt} + K_s\theta = KBI$$

The characteristic equation may be written;

$$Jp^2 + Dp + K_s = 0$$

where D = damping appropriate to the supply.
Roots of the equation are p_1 and p_2 where

$$p_1 = -\frac{D}{2J} + \sqrt{\left(\frac{D^2}{4J^2} - \frac{K_s}{J}\right)}; \; p_2 = -\frac{D}{2J} - \sqrt{\left(\frac{D^2}{4J^2} - \frac{K_s}{J}\right)}.$$

Various types of response may occur, namely under-damped, critically damped or over-damped.

In this example,

$$a = \frac{D}{2J}$$

$$\text{and} \quad \omega_d = \sqrt{\left(\frac{K_s}{J} - \frac{D^2}{4J^2}\right)}.$$

Critical damping occurs when $\dfrac{D^2}{4J^2} = \dfrac{K_s}{J}$

$$\therefore \quad D_{\text{crit}} = 2\sqrt{(K_sJ)}.$$

Also $\omega_n = \sqrt{\dfrac{K_s}{J}}$ i.e. when there is no damping.

$$\text{Damping ratio, } \zeta = \frac{D}{J} \cdot \frac{1}{2\omega_n} = \frac{D}{2\sqrt{(JK_s)}}.$$

The general statement of the solution is expressed;

$$\theta = \theta_\infty\left[1 - \frac{1}{1 - \zeta^2}\,\epsilon^{-\zeta\omega_n t}\{\sin(\sqrt{(1 - \zeta^2)} \cdot \omega_n t + \phi)\}\right]$$

where

$$\tan\phi = \frac{\sqrt{(1-\zeta^2)}}{\zeta}$$

$$\theta_\infty = \text{final steady-state value of } \theta.$$

Time-dependant forcing functions

The dynamic equilibrium equations obtained for energy conversion devices have been seen to be generally of the form of differential equations. Adoption of assumptions providing constant coefficients, and the use of the differential operator 'p', permit manipulation according to ordinary algebraic laws.

The response to a forcing function consists of a natural response, which decays with time, added to the steady-state response which depends on the steady-state nature of the forcing function. The natural response is determined by the constants of the system.

(*a*) *Consider a forcing function which is proportional to time, commonly termed a 'ramp' function.*

Principles are demonstrated for a simple series RL circuit;

We have,
$$L\frac{di}{dt} + iR = Et.$$

The natural response, as before,

$$i_n = A\epsilon^{-t/T}.$$

For a linear system the output will be of the same form as the input. Thus for the particular integral,

Let
$$i_{ss} = Bt + C$$

where B and C are constants. Substitute in the original equation,

$$LB + (Bt + C)R = Et$$

$$\text{or,} \quad TB + Bt + C = \frac{E}{R}t \text{ where } T = \frac{L}{R}.$$

Equating coefficients,

$$B = \frac{E}{R} \text{ and } C = -TB = -\frac{E}{R}T$$

$$\therefore \; i_{ss} = \frac{E}{R}t - \frac{E}{R}T.$$

Combining with the natural response,

$$i = \frac{E}{R}t - \frac{E}{R}T + A\epsilon^{-t/T}.$$

At $t = 0$ let $i = 0$,

$$\therefore \quad 0 = -\frac{E}{R}T + A \text{ and } A = \frac{E}{R}T.$$

Thus the solution is, $i = \frac{E}{R}(t - T + T\epsilon^{-t/T})$.

Note that there is a steady-state lag, Fig. 1.5.

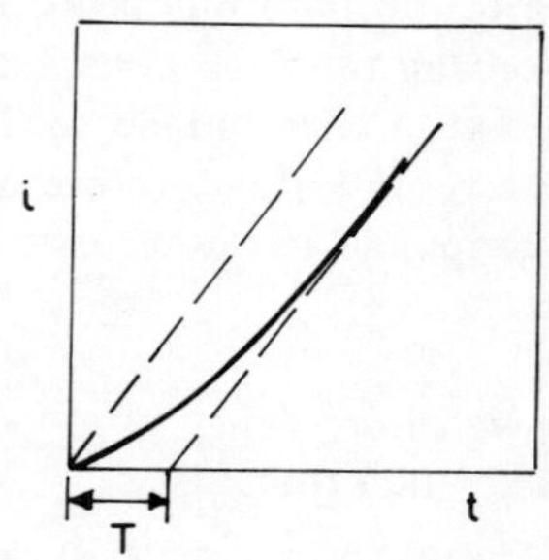

Fig. 1.5.

(*b*) *Steady-state frequency response.*

This is the study of dynamic response due to a forcing function which is time variable in steady-state. It has an important significance in the analysis of systems and devices as will be seen later.

By using a Fourier series expansion, steady-state periodic forcing functions can be represented by a sum of sinusoidal functions. Hence the steady-state response of a linear system to a periodic forcing function will be the sum of the sinusoidal responses, because the principle of superposition applies. Expressions which have been obtained for transient response require very little modification to obtain solutions for steady-state frequency response, for example;

The impedance of a series RL circuit to a current, $i = I\epsilon^{pt}$ is obtained as $e/i = Z$.

Since

$$e = L\frac{di}{dt} + iR,$$

then $Z = (Lp + R)$, which may be expressed as Z (function of p), or simply $Z(p)$.

Now if $i = I\epsilon^{j\omega t}, \ \dfrac{di}{dt} = j\omega I \epsilon^{j\omega t}$

and $e = I(j\omega L + R)\epsilon^{j\omega t}$

giving $Z(j\omega) = e/i = (j\omega L + R)$.

Hence the solution is obtained by substituting $(j\omega)$ for p.

The function $\epsilon^{j\omega t}$ can be expressed as the sum of sinusoids,

$$\epsilon^{j\omega t} = (\cos \omega t + j \sin \omega t).$$

Using this expansion above,

$$i = I\epsilon^{j\omega t} = I(\cos \omega t + j \sin \omega t)$$
$$e = IZ\epsilon^{j\omega t} = IZ(\cos \omega t + j \sin \omega t).$$

For a linear system, the real parts of e and i correspond, and likewise the imaginary or j-parts, and $Z(j\omega)$ applies equally to either part.

The general concept of frequency response embraces the case of a complex frequency, $s = (\sigma + j\omega)$, and for a function of the form ϵ^{st}, the frequency response is obtained by substituting s for p.

(c) *The Laplace transform.*

An important method of solution of systems subject to time-dependant forcing functions follows from the concept that the function, including step functions, may be interpreted as in Fourier analysis, as an integration of component steady sinusoids. The time-dependant function, $f(t)$, may thus be transformed into a function of $s = F(s)$, using the Laplace integration.

$$F(s) = \int_0^\infty f(t) \cdot \epsilon^{-st} \cdot dt$$

that is, the function to be transformed is multiplied by ϵ^{-st} where s corresponds to frequency and the product is integrated through all time. The result, $F(s)$, is a measure of the component ϵ^{st} in the function $f(t)$. (Ref. 4).

Consider the Unit Step function where the time function is simply 1.

$$F(s) = \int_0^\infty 1 \cdot \epsilon^{-st} \cdot dt = -\frac{1}{s}[\epsilon^{-st}]_0^\infty$$
$$= -\frac{1}{s}(0 - 1) = \frac{1}{s}.$$

The time function 1 and the transform $1/s$ are termed a Laplace-transform pair. Application of this procedure to various common standard forms of time functions enables a table of Transforms to be drawn up for ready reference, (p.19).

The following important example is also used to demonstrate application of the transform.

(*d*) *Transient response to an a.c. voltage.*

Consider a series RL circuit with a function of a.c. voltage, $f(t) = V_{max} \sin \omega t$, applied.

Using the complex frequency operator s, the equation becomes:

$$(Ls + R)I = F(s).$$

It is thus necessary to transform $f(t)$ into terms of s. This is achieved by using the Laplace integral,

$$\begin{aligned}
F(s) &= \int_0^\infty f(t) \cdot \epsilon^{-st}\, dt \\
&= V_{max} \int_0^\infty \sin \omega t \cdot \epsilon^{-st} \cdot dt \\
&= V_{max} \int_0^\infty \left(\frac{\epsilon^{j\omega t} - \epsilon^{-j\omega t}}{2j} \right) \cdot \epsilon^{-st} \cdot dt \\
&= \frac{V_{max}}{2j} \int_0^\infty [\epsilon^{-(s-j\omega)t} - \epsilon^{-(s+j\omega)t}]\, dt \\
&= \frac{V_{max}}{2j} \left(\frac{1}{s - j\omega} - \frac{1}{s + j\omega} \right) = \frac{V_{max}\, \omega}{s^2 + \omega^2}
\end{aligned}$$

$$\therefore \quad I = \frac{V_{max}\, \omega}{(s^2 + \omega^2)} \cdot \frac{1}{(R + Ls)}.$$

Using partial fractions;

$$I = \frac{V_{max}\, \omega L}{(R^2 + \omega^2 L^2)} \left[\frac{1}{(s + R/L)} - \frac{s}{(s^2 + \omega^2)} + \frac{R/L}{(s^2 + \omega^2)} \right].$$

Transforming back to terms of t,

$$\begin{aligned}
i(t) &= \frac{V_{max}\, \omega L}{(R^2 + \omega^2 L^2)} \left[\epsilon^{-Rt/L} - \cos \omega t + \frac{R}{\omega L} \sin \omega t \right] \\
&= V_{max} \left[\frac{\omega L}{R^2 + \omega^2 L^2}\, \epsilon^{-Rt/L} + \frac{1}{(R^2 + \omega^2 L^2)^{\frac{1}{2}}} (\sin \omega t \cos \phi - \cos \omega t \sin \phi) \right.
\end{aligned}$$

$$= V_{max}\left[\frac{\omega L}{R^2 + \omega^2 L^2}\,\epsilon^{-Rt/L} + \frac{1}{(R^2+\omega^2L^2)^{\frac{1}{2}}}\sin(\omega t - \phi)\right]$$

where $\tan\phi = \omega L/R$.

This solution may be expressed in other forms, viz;

$$i(t) = \frac{V_{max}}{(R^2+\omega^2L^2)^{\frac{1}{2}}}\left[\frac{\omega L}{(R^2+\omega^2L^2)^{\frac{1}{2}}}\epsilon^{-t/T} + \sin(\omega t - \phi)\right]$$

$$= I_{max}\,[\sin\phi \cdot \epsilon^{-t/T} + \sin(\omega t - \phi)]$$

$=$ Transient (d.c.) part + Final steady-state part.

Note that this solution is the special case for switching at $t = 0$, i.e. the instant when the applied voltage is at zero. More generally:

$$i = I_{max}\sin(\omega t - \phi) + A\epsilon^{-t/T}.$$

Then, if switching occurs at time t' sec after zero in the voltage cycle, at switching since $i = 0$,

$$i = 0 = I_{max}\sin(\omega t' - \phi) + A\epsilon^{-t'/T}.$$

$$\therefore\; A = -I_{max}\sin(\omega t' - \phi)\,\epsilon^{+t'/T}.$$

Hence, in general,

$$i = I_{max}\,[\sin(\omega t - \phi) - \sin(\omega t' - \phi)\,\epsilon^{(t'-t)/T}].$$

The exponential term represents a transient current component which produces asymmetry in the initial response current. For the condition $t' = t = 0$, a 'doubling' of the initial current occurs. If $\omega t' = \phi$, no transient effect occurs.

These phenomena are very important in all switching operations, including sudden short-circuits, related to machines or other reactive devices.

N.B. The following Laplace transform pairs are in fairly common usage:

$f(t) =$	V	$V\epsilon^{-at}$	$V\sin\omega t$	$V\cos\omega t$
$F(s) =$	$\frac{V}{s}$	$\frac{V}{s+a}$	$\frac{V\omega}{s^2+\omega^2}$	$\frac{Vs}{s^2+\omega^2}$

Magnetic coupling

Consider two stationary coils which share a common magnetic

circuit and are supplied with currents i_1 and i_2 respectively, magnetising in the same sense, Fig. 1.6.

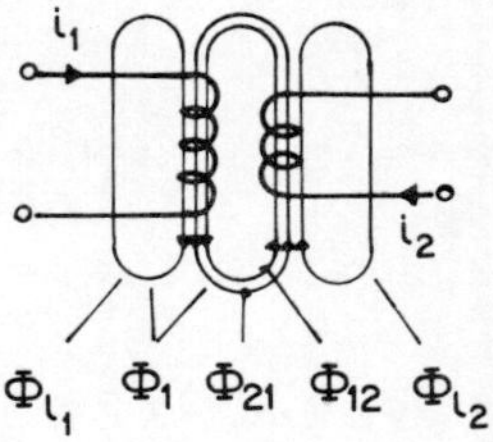

Fig. 1.6.

A linear m.m.f./flux relation is assumed.

Let i_1 cause a flux Φ_1 in coil 1, and

$$\Phi_1 = \Phi_{21} + \Phi_{l_1}$$

where Φ_{21} = flux due to i_1 linking coil 2, termed magnetising flux

Φ_{l_1} = flux due to i_1 not linking coil 2 and termed leakage flux.

Similarly, $\Phi_2 = \Phi_{12} + \Phi_{l2}$ in coil 2 due to i_2.

Flux linkages for the coils are,

$$\lambda_1 = N_1 \Phi_1 + N_1 \Phi_{12}$$

$$\lambda_2 = N_2 \Phi_2 + N_2 \Phi_{21}.$$

Now self inductance = self flux-linkage per amp of self current

and $$L_{11} = \frac{N_1 \Phi_1}{i_1} = \frac{N_1 (\Phi_{21} + \Phi_{l_1})}{i_1}$$

$$= L_{m_1} + L_{l_1}$$

where L_{m_1} = magnetising inductance

L_{l_1} = leakage inductance.

Also $$L_{22} = \frac{N_2 \Phi_2}{i_2} = \frac{N_2 (\Phi_{12} + \Phi_{l2})}{i_2}$$

$$= L_{m2} + L_{l2}.$$

Mutual fluxes Φ_{21} and Φ_{12} thread a common path of permeance $\mathcal{P}$.

Then,

$$\Phi_{21} = \mathcal{P} N_1 i_1 ; \; \Phi_{12} = \mathcal{P} N_2 i_2 .$$

Hence the mutual inductances are, in coil 1 due to coil 2,

$$L_{12} = \frac{N_1 \Phi_{12}}{i_2} = \mathcal{P} N_1 N_2$$

In coil 2 due to coil 1,

$$L_{21} = \frac{N_2 \Phi_{21}}{i_1} = \mathcal{P} N_1 N_2$$

$$\therefore \quad L_{12} = L_{21} = M.$$

Now
$$\begin{aligned} \lambda_1 &= N_1 \Phi_1 + N_1 \Phi_{12} \\ &= L_{11} i_1 + (N_1 \mathcal{P} N_2 i_2) \\ &= L_{11} i_1 + M i_2 \end{aligned}$$

Similarly
$$\lambda_2 = L_{22} i_2 + M i_1 .$$

Then the voltages induced in the coils are,

$$e_1 = -\frac{d\lambda_1}{dt} = -\left(L_{11} \frac{di_1}{dt} + M \frac{di_2}{dt}\right)$$

$$e_2 = -\frac{d\lambda_2}{dt} = -\left(L_{22} \frac{di_2}{dt} + M \frac{di_1}{dt}\right).$$

A *coupling factor*, k_1, defines the effectiveness of coil 1 in producing mutual magnetising flux, viz;

$$k_1 = \frac{L_{m1}}{L_{11}} = \frac{N_1 M}{N_2 L_{11}} .$$

Similarly,
$$k_2 = \frac{L_{m2}}{L_{22}} = \frac{N_2 M}{N_1 L_{22}} .$$

The *coefficient of coupling*, *k*, is the geometric mean of k_1 and k_2, viz;

$$k = \sqrt{(k_1 k_2)} = \frac{M}{\sqrt{(L_{11} L_{22})}}$$

$$\text{or,} \quad M = k\sqrt{(L_{11} L_{22})}$$

and k relates Mutual to Self inductances.

The leakage coefficient

$$\sigma = (1 - k^2) = 1 - \frac{M^2}{L_{11} L_{22}}$$

and with perfect coupling $k = 1$ and $\sigma = 0$.

Transients in coupled circuits

Transient phenomena may arise due to sudden short circuit of one coil, or due to sudden application of voltage to one coil when the other coil is either on open circuit or short-circuited by a low resistance. These conditions can arise in field or armature circuits of machines, the presence of damping windings being an added complication.

(*a*) *Effective inductance of coupled coils.*

Consider two coupled coils, one being short-circuited by a low resistance. Neglecting resistance, the voltage equations can be written,

$$v_1 = L_1 \frac{di_1}{dt} + M \frac{di_2}{dt}$$

$$v_2 = L_2 \frac{di_2}{dt} + M \frac{di_1}{dt} = 0 \text{ due to s/c.}$$

$$\therefore \quad \frac{di_2}{dt} = -\frac{M\,di_1}{L_2 dt}.$$

Substituting in the expression for v_1,

$$v_1 = L_1 \frac{di_1}{dt} - \frac{M^2\,di_1}{L_2\,dt}$$

$$= L_1 (1 - k^2) \frac{di_1}{dt}.$$

The result, due to the s/c coupled coil, (neglecting resistance), is that the effective self inductance and the effective time constant of the circuit are reduced. Thus,

$$L_1' = L_1(1 - k^2)$$

$$T_1' = L_1'/R_1.$$

(*b*) *Transient response.*

With resistance included, the voltage equations are,

$$v_1 = i_1 R_1 + L_1 \frac{di_1}{dt} + M \frac{di_2}{dt}$$

$$v_2 = i_2 R_2 + L_2 \frac{di_2}{dt} + M \frac{di_1}{dt}.$$

(*i*) with the secondary coil on open-circuit,

$$i_2 = 0$$

$$v_1 = i_1 R_1 + L_1 \frac{di_1}{dt}$$

$$v_2 = M \frac{di_1}{dt}$$

Applying a step function of d.c. voltage V,

$$i_1 = \frac{V}{(R_1 + L_1 p)} = \frac{V}{R_1 (1 + T_1 p)}$$

$$\therefore \quad i_1 = \frac{V}{R_1} (1 - \epsilon^{-t/T_1}) \text{ where } T_1 = L_1/R_1$$

$$v_2 = \frac{MV}{T_1 R_1} \epsilon^{-t/T_1}$$

$$= \frac{MV}{L_1} \epsilon^{-t/T_1}$$

$$= Vk (L_2/L_1)^{\frac{1}{2}} \epsilon^{-t/T_1} .$$

The nature of the secondary response depends on the ratio L_1/L_2. If $L_1 \ll L_2$ the response approaches an impulse.

(*ii*) with the secondary coil on short-circuit.

The voltage equations are,

$$v_1 = i_1 R_1 + L_1 \frac{di_1}{dt} + M \frac{di_2}{dt}$$

$$v_2 = i_2 R_2 + L_2 \frac{di_2}{dt} + M \frac{di_1}{dt} = 0 \text{ on s/c.}$$

Applying a step function of d.c. voltage V.

The steady-state currents are,

$$I_{1(ss)} = \frac{V}{R_1}; \; I_{2(ss)} = 0.$$

The transient currents are of the form,

$$i_1 = I_1 \epsilon^{pt} \text{ and } i_2 = I_2 \epsilon^{pt} .$$

Substituting in the characteristic equations gives,

$$0 = (R_1 + L_1 p) I_1 + Mp I_2$$

$$0 = (R_2 + L_2 p) I_2 + Mp I_1 .$$

Eliminating the currents, therefore,

$$(R_1 + L_1p)(R_2 + L_2p) - M^2p^2 = 0$$

$$(L_1L_2 - M^2)p^2 + (R_1L_2 + R_2L_1)p + R_1R_2 = 0.$$

Now
$$\frac{L_1L_2 - M^2}{L_1L_2} = 1 - \frac{M^2}{L_1L_2}$$

$$= (1 - k^2) = \sigma$$

$$\therefore \quad \sigma(L_1L_2)p^2 + (R_1L_2 + R_2L_1)p + R_1R_2 = 0.$$

Dividing across by $R_1 R_2$ gives,

$$\sigma T_1 T_2 p^2 + (T_1 + T_2)p + 1 = 0$$

where $T_1 = L_1/R_1$ and $T_2 = L_2/R_2$.

Hence the roots are,

$$p_1, p_2 = \frac{-(T_1 + T_2) \pm [(T_1 + T_2)^2 - 4\sigma T_1 T_2]^{\frac{1}{2}}}{2\sigma T_1 T_2}.$$

Now the solution for current is of the form,

$$i_1 = I_{1(ss)} + A\epsilon^{p_1t} + B\epsilon^{p_2t}.$$

To solve for A and B we require another equation, thus,

$$\frac{\mathrm{d}i_1}{\mathrm{d}t} = p_1 A\epsilon^{p_1t} + p_2 B\epsilon^{p_2t}.$$

A and B are obtained by substituting known initial conditions. Thus at $t = 0$;

$$i_{10} = 0; \; i_{20} = 0.$$

Substituting in the original voltage equations, the iR terms are zero and,

$$V = \left(L_1 - \frac{M^2}{L_2}\right)\frac{\mathrm{d}i_1}{\mathrm{d}t_0}$$

$$= \sigma L_1 \frac{\mathrm{d}i_1}{\mathrm{d}t_0}$$

$$\therefore \quad \frac{\mathrm{d}i_1}{\mathrm{d}t_0} = \frac{V}{\sigma L_1}.$$

Also,
$$\frac{\mathrm{d}i_2}{\mathrm{d}t_0} = -\frac{M}{L_2} \cdot \frac{\mathrm{d}i_1}{\mathrm{d}t_0} = -\frac{M}{L_2}\frac{V}{\sigma L_1}$$

which we shall require to determine i_2.

Hence we have,

$$0 = \frac{V}{R_1} + A + B$$

$$\frac{V}{\sigma L_1} = p_1 A + p_2 B$$

and A and B are evaluated by solving these two equations.

Further, since $$i_2 = I_{2(ss)} + C\epsilon^{p_1 t} + D\epsilon^{p_2 t}$$

the constants C and D are evaluated using similar reasoning.

Approximations:

(1) If $\sigma \to 0$, i.e. negligible leakage, then the roots p_1 and p_2 may be approximated as follows;

In the quadratic for p the p^2 term becomes negligible and,

$$p_1 = \frac{-1}{(T_1 + T_2)}.$$

The second root is obtained by neglecting the $4\sigma T_1 T_2$ term under the radical,

and $$p_2 = -\frac{1}{\sigma}\left(\frac{1}{T_1} + \frac{1}{T_2}\right).$$

Note also that the main delay is then mainly due to the p_1 term, that is, the effective time constant $\approx (T_1 + T_2)$.

(2) If resistance may be neglected a simpler solution occurs since effective time constant approaches $T_1' = \sigma T_1$, and,

$$i_1 = I_{1(ss)}\,(1 - \epsilon^{-t/T_1'}).$$

REFERENCES

1. HAM, J.M. and SLEMON, G.R. (1961). *The Scientific Basis of Electrical Engineering*, Wiley, New York.
2. HAMMOND, P. *Electromagnetism for Engineers*, Pergamon Press, Oxford.
3. HAMMOND, S.B. (1961). *Electrical Engineering*, McGraw-Hill, New York.
4. GUILLEMIN, E.A. (1953). *Introductory Circuit Theory*. Wiley, New York.

EXAMPLES

EXAMPLE 1.1.

A coil having an inductance of 4H and resistance of 10 Ω is switched across 20V d.c. Calculate (a) the time for the current to reach 50% of its final value, (b) the instantaneous value of the current after 0·5 sec.

Solution:

$$L\,di/dt + iR = E.$$

Then $i = I(1 - \epsilon^{-t/T})$ where $T = L/R$.

(a) This may be written,

$$Rt/L = l_n \frac{I}{(I - i)}$$

$$\therefore \; t = \frac{L}{R} l_n \frac{I}{(I - i)}$$

Now $I = E/R = 20/10 = 2\text{A}$

$$\therefore \; t = \frac{4}{10} l_n \frac{2}{(2 - 1)} = \frac{4}{10} \times 0{\cdot}693$$

$$= 0{\cdot}278 \text{ sec.}$$

(b) $i = I(1 - \epsilon^{-t/T})$

$$i_{0\cdot5} = 2(1 - \epsilon^{-0\cdot5/T}) = 2(1 - \epsilon^{-1\cdot25})$$

$$= 1{\cdot}43\text{A}.$$

EXAMPLE 1.2.

A straight conductor operates as an elementary linear motor in the horizontal plane, on parallel conducting rails 1 m apart. The magnetic field is vertical and has a uniform density of 0·5Wb/m². Viscous friction between the conductor and the rails is 0·75N sec/m, and the mass of the moving part is 0·5 kg. The conductor has ohmic resistance of 1Ω and the potential applied between the rails is 100V, d.c. Derive an expression for the instantaneous velocity of the conductor and hence calculate, (a) the steady-state velocity, (b) the steady-state current in the conductor, (c) the instantaneous acceleration and the time from rest when the instantaneous velocity is 24 m/sec.

Solution:

Summing the forces acting on the conductor,

$$M\frac{dv}{dt} + F_v \cdot v = Bli$$

$$M\frac{dv}{dt} + F_v \cdot v = \frac{Bl(E - Blv)}{R}$$

$$M\frac{dv}{dt} + \left(F_v + \frac{B^2l^2}{R}\right)v = \frac{EBl}{R}$$

$$v_{ss} = \frac{EBl}{R(F_v + B^2l^2/R)}$$

and $v = v_{ss}(1 - \epsilon^{-t/T})$ where $T = \dfrac{M}{(F_v + B^2l^2/R)}$.

(a) $$v_{ss} = \frac{100 \times 0{\cdot}5 \times 1}{\left(0{\cdot}75 + \dfrac{0{\cdot}5^2 \times 1}{1}\right)}$$

$$= \frac{50}{(0{\cdot}75 + 0{\cdot}25)} = 50 \text{ m/sec.}$$

(b) $$I_{ss} = \frac{E - Blv_{ss}}{R} = \frac{100 - 25}{1}$$

$$= 75\text{A.}$$

(c) $$M\frac{dv}{dt} + (0{\cdot}75 \times 24) = \frac{EB - B^2l^2v}{R}$$

$$= 50 - 6$$

$$M\frac{dv}{dt} = 26$$

$$\therefore \quad \frac{dv}{dt} = 52 \text{ m/sec/sec.}$$

$$v = 50(1 - \epsilon^{-2t})$$

$$\frac{26}{50} = \epsilon^{-2t}$$

$$= 0{\cdot}52$$

$$\therefore \quad t = 0{\cdot}325 \text{ sec.}$$

EXAMPLE 1.3.

A single-layer coil of 50 turns and resistance 6Ω is closely wound on a thin cylinder of diameter 2 cm, forming a mass of 5 g. The coil is located concentrically in a radial magnetic field at a uniform flux density of 0·8 Wb/m². Current is supplied to the coil from, (a) a constant-current source, (b) a constant-voltage source. Calculate, in each case, the instantaneous axial velocity of the coil per unit of supply if frictional damping is 0·3 N sec/m.

An annular diaphragm constraint having a spring constant of 5000 N/m is now attached to the coil. Calculate the undamped natural frequency and the damping ratio, for the constant-current supply, assuming the mass and the damping remain unchanged.

Solution:

$$\text{Total length of conductor, } l = 50 \times \pi \times 2 \times 10^{-2} = \pi \text{ m.}$$

The conductor and the field are at right angles,

$$\therefore \quad f = Bli = 0{\cdot}8\pi \text{ N/A.}$$

Case (a) $M\,dv/dt + F_v v = f$

$$v_{ss}, \text{ (when } dv/dt = 0), = \frac{f}{F_v} = \frac{0{\cdot}8\pi}{0{\cdot}3} = 8{\cdot}4 \text{ m/sec/A.}$$

Considering time-constant, T,

$$\frac{1}{T} = \frac{F_v}{M} = \frac{0{\cdot}3}{5 \times 10^{-3}} = 60$$

$$\therefore \quad v = 8{\cdot}4\,(1 - \epsilon^{-60t})\text{ m/sec/A.}$$

Case (b)

$$M\,dv/dt + F_v v = Bli$$

$$= Bl\left(\frac{E - Blv}{R}\right) = \frac{EBl}{R} - \frac{B^2l^2v}{R}$$

$$\therefore \quad M\,dv/dt + \left(F_v + \frac{B^2l^2}{R}\right)v = \frac{EBl}{R}$$

$$v_{ss} = \frac{Bl}{R}\cdot\frac{1}{\left(F_v + \frac{B^2l^2}{R}\right)} = \frac{0{\cdot}8\pi}{6}\cdot\frac{1}{(0{\cdot}3 + 1{\cdot}06)}$$

$$= \frac{0{\cdot}8\pi}{8{\cdot}15} = 0{\cdot}31 \text{ m/sec/V}$$

$$\frac{1}{T} = \frac{\text{Damping}}{M} = \frac{1{\cdot}36 \times 10^3}{5} = 272$$

$$\therefore \quad v = 0{\cdot}31\,(1 - \epsilon^{-272t})\text{ m/sec/V.}$$

Last part: Let θ = instantaneous axial linear displacement.

Then $\quad M\,\mathrm{d}^2\theta/\mathrm{d}t^2 + F_v \cdot \mathrm{d}\theta/\mathrm{d}t + K\theta = f.$

This is a second order equation and hence,

$$\omega_n = (K/M)^{\frac{1}{2}} = (5000/5 \times 10^{-3})^{\frac{1}{2}}$$

$$= 10^3 = 2\pi f_n$$

$$\therefore f_n = 159\,\mathrm{c/s}$$

$$a = \frac{F_v}{2M} = \frac{0{\cdot}3}{10 \times 10^{-3}} = 30$$

$$\therefore \zeta = \frac{a}{\omega_n} = \frac{30}{1000} = 0{\cdot}03.$$

EXAMPLE 1.4.

Determine the Laplace transform of the following time functions; (a) $V\epsilon^{-at}$, (b) $V\cos\omega t$, (c) $V\cos(\omega t + \phi)$ (d) $V\epsilon^{-at}\sin\omega t$.

Solution:

The procedure in each case involves applying the Laplace integral, since

$$F(s) = \int_0^\infty f(t) \cdot \epsilon^{-st}\,\mathrm{d}t$$

$$\text{(a) } F(s) = V\int_0^\infty \epsilon^{-at}\,\epsilon^{-st}\,\mathrm{d}t$$

$$= V\int_0^\infty \epsilon^{-(s+a)t}\,\mathrm{d}t$$

$$= \frac{V}{(s + a)}.$$

$$\text{(b) } F(s) = V\int_0^\infty \cos\omega t \cdot \epsilon^{-st}\,\mathrm{d}t$$

$$= \frac{V}{2}\int_0^\infty (\epsilon^{j\omega t} + \epsilon^{-j\omega t})\,\epsilon^{-st}\,\mathrm{d}t$$

$$= \frac{V}{2}\int_0^\infty \epsilon^{-(s-j\omega)t} + \epsilon^{-(s+j\omega)t}\,\mathrm{d}t$$

$$= \frac{V}{2}\left(\frac{1}{s - j\omega} + \frac{1}{s + j\omega}\right)$$

$$= \frac{Vs}{s^2 + \omega^2}.$$

$$\text{(c) } f(t) = V\cos(\omega t + \phi)$$

$$= V(\cos\omega t\cos\phi - \sin\omega t\sin\phi)$$

$$F(s) = \int_0^\infty V\cos(\omega t + \phi)\cdot\epsilon^{-st}\,\mathrm{d}t$$

$$= V[\cos\phi\int_0^\infty \cos\omega t\cdot\epsilon^{-st}\,\mathrm{d}t - \sin\phi\int_0^\infty \sin\omega t\cdot\epsilon^{-st}\,\mathrm{d}t]$$

$$= V\left[\frac{(\cos\phi)s}{s^2+\omega^2} - \frac{(\sin\phi)\omega}{s^2+\omega^2}\right]$$

Hence

$$F(s) = V\left[\frac{(\cos\phi)s - (\sin\phi)\omega}{(s^2+\omega^2)}\right].$$

(d) $f(t) = V\epsilon^{-at}\sin\omega t$, and is a damped sinusoid.

$$F(s) = V\int_0^\infty \epsilon^{-at}\sin\omega t\cdot\epsilon^{-st}\,\mathrm{d}t$$

$$= \frac{V}{2j}\int_0^\infty (\epsilon^{j\omega t} - \epsilon^{-j\omega t})\,\epsilon^{-(s+a)t}\cdot\mathrm{d}t$$

$$= \frac{V}{2j}\int_0^\infty [\epsilon^{-(s+a-j\omega)t} - \epsilon^{-(s+a+j\omega)t}]\,\mathrm{d}t$$

$$= V/2j\,[1/(s+a-j\omega) - 1/(s+a+j\omega)]$$

$$= V\omega/[(s+a)^2+\omega^2].$$

EXAMPLE 1.5.

Derive an expression for the instantaneous current in a series R, L circuit due to the application of a voltage which is proportional to time, (i.e. a ramp function).

Solution:

The following solution offers an alternative method of solution of the problem discussed on p.15 .

In this case the voltage as a time function $= Et$. Using Laplace transforms,

$$F(s) = E\int_0^\infty t\cdot\epsilon^{-st}\cdot\mathrm{d}t.$$

We integrate by parts, $\left[\text{viz. } \int uv\cdot\mathrm{d}x = u\int v\cdot\mathrm{d}x - \int\left(\frac{\mathrm{d}u}{\mathrm{d}x}\int v\cdot\mathrm{d}x\right)\cdot\mathrm{d}x\right]$

$$\therefore\ F(s) = E\int_0^\infty t\cdot\epsilon^{-st}\cdot\mathrm{d}t$$

$$= E\left\{\left[\frac{t\cdot\epsilon^{-st}}{-s}\right]_0^\infty - \int_0^\infty \frac{\epsilon^{-st}}{-s}\cdot\mathrm{d}t\right\}$$

$$= E\left\{0 + \frac{1}{s^2}\right\} = \frac{E}{s^2}.$$

Now, $(Ls + R)I = \frac{E}{s^2}$

$$\therefore\ I = \frac{E}{(Ls + R)s^2} = \frac{E}{L}\left[\frac{1}{(s + R/L)s^2}\right]$$

Using partial fractions,

$$= \frac{E}{L}\left[\frac{L/R}{(s + R/L)} + \frac{1}{s^2} - \frac{L/R}{s}\right].$$

Transforming back to functions of time,

$$\therefore\ i = \frac{E}{R}[T\epsilon^{-t/T} + t - T]$$

which is the same as obtained previously.

2 Analogues

Analogues and duals

When two physical systems can be described in terms of equations of the same mathematical form, they are said to be 'analogues' of each other. The analogy is extended to the individual elements of the system, and serves to emphasise the dynamical similarity.

Hence it is possible, by choosing appropriate scale factors, to set up, for example, an electrical 'model' or 'equivalent circuit' of a mechanical system. The convenience in testing and the high accuracy obtainable using electrical components and measurements make an equivalent electrical circuit an important tool in the analysis of the system. Conversely, it can be claimed that the nature of certain electrical phenomena may be better understood if interpreted in terms of analogous mechanical devices.

The equation of a simple mechanical system can be written, in terms of the forces and the dependant variable velocity, v; Fig. 2.1(a).

$$M\frac{\mathrm{d}v}{\mathrm{d}t} + F_v \cdot v + K\int v \cdot \mathrm{d}t = F.$$

A similar mathematical form occurs for an electrical circuit, in terms of voltages expressed as functions of current, i; Fig. 2.1(b).

$$L\frac{\mathrm{d}i}{\mathrm{d}t} + Ri + \frac{1}{C}\int i \cdot \mathrm{d}t = E.$$

We can interpret

$$v = \frac{\mathrm{d}x}{\mathrm{d}t} \text{ and } i = \frac{\mathrm{d}q}{\mathrm{d}t}$$

and rewrite the above equations,

$$M\frac{d^2x}{dt^2} + F_v \cdot \frac{dx}{dt} + Kx = F$$

$$L\frac{d^2q}{dt^2} + R \cdot \frac{dq}{dt} + \frac{1}{C} \cdot q = E.$$

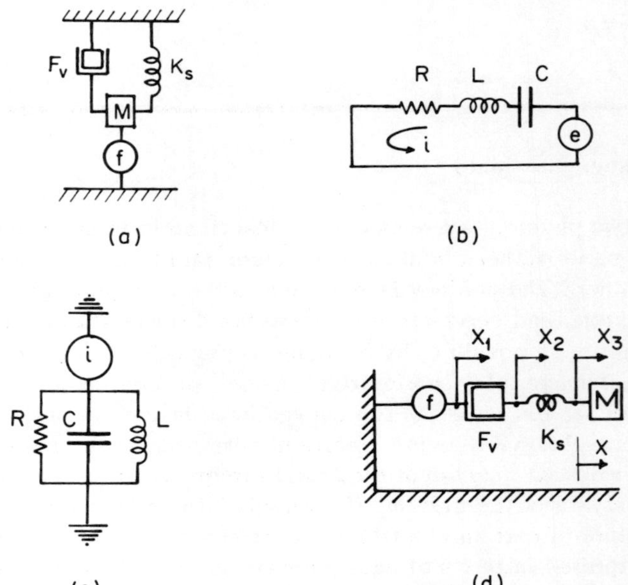

Fig. 2.1.

Comparison of the mathematical forms reveals the 'Mass/Inductanc analogue wherein mass and inductance, velocity and current, viscous friction and resistance, etc. are analogous quantities. These are summarised in the table below.

Analogy between systems belonging to the same physical group is termed 'duality' and the systems are termed 'duals' of each other. Duality does not imply equivalence, but only that the associated mathematical form is similar. It can arise from writing 'loop' equations for one system and 'node' equations for the other, e.g. the electrical case cited above applies to the series circuit and is obtained by adding voltages around a loop. The equation for the parallel electrical circuit provides the 'dual', and is obtained by adding current components at a node or junction, viz; Fig. 2.1 (c).

$$C\frac{de}{dt} + \frac{1}{R}\cdot e + \frac{1}{L}\int e\cdot dt = I.$$

This system may be used as the analogue of the mechanical case and provides an alternative, or the 'Mass/Capacitance' analogue.

We can include the mechanical torsional case, viz;

$$J\frac{d\omega}{dt} + F_{\omega}\cdot\omega + K\int\omega\cdot dt = \textcircled{T}$$

and hence tabulate analogous quantities as follows:

Mechanical		Electrical	
Linear	Torsional	Mass/Inductance	Mass/Capacit.
F	Ⓣ	E	I
M	J	L	C
v	ω	i	e
F_v	F_{ω}	R	$1/R$
$1/K$	$1/K$	C	L
x	θ	q	$\int e\cdot dt$

Further examination of the individual elements, which apparently become analogous following the preceding mathematical comparison, adds further to the dynamic similarity which can be seen to exist between mechanical and electrical elements and systems.

In the mechanical system, energy stored in the mass is given by the kinetic energy, $\frac{1}{2}Mv^2$. Energy storage occurs in a spring from the displacement, $x = \int v\cdot dt$, due to a force. This force is expressed in terms of a spring constant, K, as $F = Kx = K\int v\cdot dt$. Assuming a linear force/displacement relation, the work done is given by the area under the graph

$$= \frac{1}{2}Fx = \frac{1}{2}\frac{1}{K}\cdot F^2.$$

A viscous friction element, such as a dashpot, is an energy dissipating element.

In the mass/inductance analogy, the inductance has stored electromagnetic energy $= \frac{1}{2}Li^2$. Application of a voltage to a capacitance causes a proportional storage of charge, $q = \int i \cdot dt$, and the voltage component is $1/C \int i \cdot dt$.

Similarly, in the mass/capacitance analogy, the energy stored in a capacitance is $\frac{1}{2} Ce^2$. The energy storage of a spring, $\frac{1}{2} (1/K) \cdot F^2$ is analogous to that in inductance, $\frac{1}{2} Li^2$ whence inductance becomes the analogue of spring compliance, $1/K$.

Scale factors

To represent mechanical quantities in terms of analogous electrical quantities, convenient sizes must be sought for the electrical quantities, particularly if it is intended to set up the analogue for experimental study. But an initial decision, say to represent 1 kg mass by 1 μF, means that the other quantities must be consistent. This involves the calculation of appropriate scale factors.

The primary measurements in a mechanical system are mass, length and time, or simply M, l and t_m. The corresponding analogous measures in the mass/inductance analogy are L, q and t_e. An initial choice of scale factors k_1, k_2 and k_3 is made, commonly dictated by practical considerations. Hence we can write:

$$L = k_1 M, \quad q = k_2 l, \quad t_e = k_3 t_m.$$

The other measurements of importance are related to these primary quantities and can now be determined as follows:

Analogy	Scale factor	Dimensions	Equivalent factor
i/velocity	$i = k_4 v$	$v = [lt_m^{-1}]$	$k_4 = \frac{k_2}{k_3}$
e/force	$e = k_5 f$	$f = [Mlt_m^{-2}]$	$k_5 = \frac{k_1 k_2}{k_3^2}$
R/friction	$R = k_6 F_v$	$F_v = \left[Mlt_m^{-2} \cdot \frac{t_m}{l}\right]$	$k_6 = \frac{k_1}{k_3}$
$C/\frac{1}{K}$	$C = k_7 \cdot \frac{1}{K}$	$\frac{1}{K} = \left[\frac{l}{Mlt_m^{-2}}\right]$	$k_7 = \frac{k_3^2}{k_1}$
ω_e/ω_m	$\omega_e = k_8 \omega_m$	$\omega_m = [t_m^{-1}]$	$k_8 = \frac{1}{k_3}$

If the resulting magnitudes obtained for the several electrical quantities are not practical, the chosen values of k_1, k_2 or k_3 may be changed. Note particularly that the time scale can be altered, sometimes with considerable advantage.

By making the appropriate changes in the first two columns, this table applies equally for the mass/capacitance analogy.

Loops and nodes

It has been pointed out that the equation for the series electrical circuit arises from the summation of voltages in a loop. Similarly the equilibrium equation for the parallel circuit is obtained by summing current components at a node.

Consider the following mechanical system, (Fig. 2.1 (c)) where x_1, x_2 and x_3 are component displacements. Force is common and the resultant displacement and velocity is the algebraic sum of the components.

Then $$x = x_1 + (x_2 - x_1) + (x_3 - x_2).$$

In terms of velocities,

$$v = v_K + v_F + v_M.$$

Now $$f = K\int v \cdot dt \quad \therefore \quad v_K = \frac{1}{K} \cdot \frac{df}{dt}$$

$$f = F_v \cdot v \quad \therefore \quad v_F = \frac{f}{F_v}$$

$$f = M \cdot \frac{dv}{dt} \quad \therefore \quad v_M = \frac{1}{M}\int f \cdot dt.$$

Hence $$\underline{\frac{1}{K} \cdot \frac{df}{dt} + \frac{f}{F_v} + \frac{1}{M}\int f \cdot dt = v.}$$

This system may be viewed as a 'series' mechanical system and is the dual of the mechanical system formerly considered, the former being thus a 'parallel' mechanical system, (Fig. 2.1 (a)), where force components are summed at a node.

If nodes and loops are recognised in this way, analogy between loop, (series), systems gives the M/C analogue. So also will analogy between node, (parallel), systems. But analogy between a node system and a loop system will give the M/L analogue.

The M/C analogue, where force and current, velocity and voltage are analogous, is commonly considered the 'direct' analogue, whereas the M/L analogy is thought of as the 'indirect' analogue. This concept of a 'direct' analogy is supported by the fact that force is a function of current, ($f = Bli$), and voltage a function of velocity, ($e = Blv$), in electromagnetism.

(This viewpoint is opposite to the 'classical' idea that voltage and force are analogous, which tenet reverses the above concepts of 'direct' and 'indirect' analogies.)

Lumped systems having more than one degree of freedom

The preceding ideas concerning the derivation of analogues can be applied to mechanical systems of a more complex nature.

There will be a condition of equilibrium for each node in a mechanical system, and hence, for each, an equation of equilibrium. Using the direct analogue there will be an electrical analogue having the same number of nodes; using the indirect analogy the electrical circuit will have a corresponding number of loops.

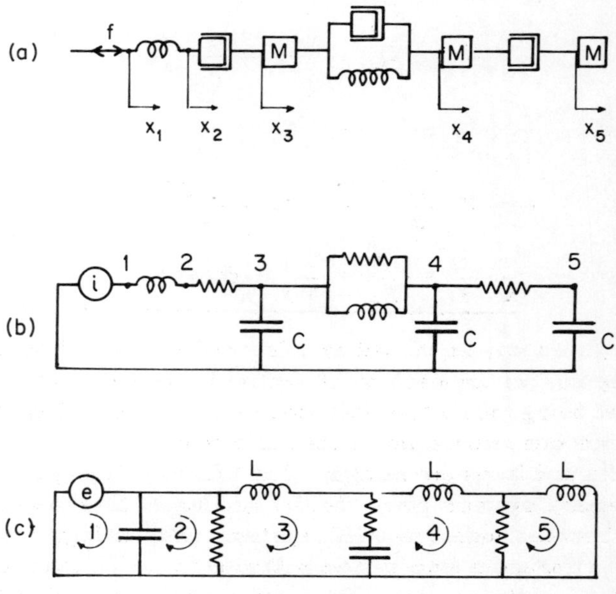

Fig. 2.2.

To demonstrate the derivation of the electrical analogue in a general case, the following example consists of masses, weightless springs and dashpots arranged in series and parallel groupings. Fig. 2.2(a). Each mass is a node or junction and in addition points 1 and 2 are also nodes. These nodes will be displaced as a result of applied force, F.

Using the direct analogy, a very similar electrical circuit is obtained. Fig. 2.2(b). Current is analogous to force and K, F_v and M are replaced by L, R and C respectively.

Using the indirect analogy, an electrical circuit is obtained having a loop for each mechanical node point. The appropriate analogous quantities occur as shown (Fig. 2.2(c)). Hence there are 5 node points in the M/C analogue and 5 loops in the M/L analogue in this example.

Equivalent circuit of an electromechanical transducer

The analogue principle is a valuable aid to the study of the dynamics of systems in which electrical and mechanical parts are directly inter-related. A total equivalent circuit can be derived which permits significant experimental and analytical study of the system. Principles are demonstrated here for the case of a moving-coil device such as an electromagnetic loudspeaker or a vibrator, in which a coil of finite mass is mechanically constrained and moves in a uniform magnetic field due to electromagnetic forces, Fig. 2.3(a).

The equivalent circuit

A unique functional relationship exists between the electrical and mechanical variables, viz; the mechanical force may be expressed $f = Bli = K_0 i$, and the induced voltage in the electrical part is a function of velocity, say $e = Blv = K_c v$ and K_0 is termed the 'transduction factor'.

Dynamical equations can be applied as follows:

For the electrical part,

$$L_e \frac{\mathrm{d}i}{\mathrm{d}t} + iR_e + e_g = E.$$

For the mechanical part,

$$M \frac{\mathrm{d}v}{\mathrm{d}t} + vF_v + K_s \int v \cdot \mathrm{d}t = K_0 i$$

where F_v = viscous friction and K_s = spring constant.

We can write, $v = \frac{1}{K_0} e_g$; $\frac{dv}{dt} = \frac{1}{K_0}\frac{de_g}{dt}$; $\int v \cdot dt = \frac{1}{K_0}\int e_g \cdot dt.$

Direct substitution in the equation for the mechanical part gives,

$$\frac{M}{K_0^2}\frac{de_g}{dt} + \frac{F_v}{K_0^2} e_g + \frac{K_s}{K_0^2} e_g \cdot dt = i.$$

Electrically this is a node system corresponding to summation of current components and it is analogous to an electrical circuit of R, L and C in parallel, the equation being,

$$C\frac{de}{dt} + \frac{e}{R} + \frac{1}{L}\int e \cdot dt = i.$$

Hence a 'total' equivalent circuit may be drawn, Fig. 2.3 (b), where $C = \frac{M}{K_0^2}$; $L = \frac{K_0^2}{K_s}$ and $R = \frac{K_0^2}{F_v}$.

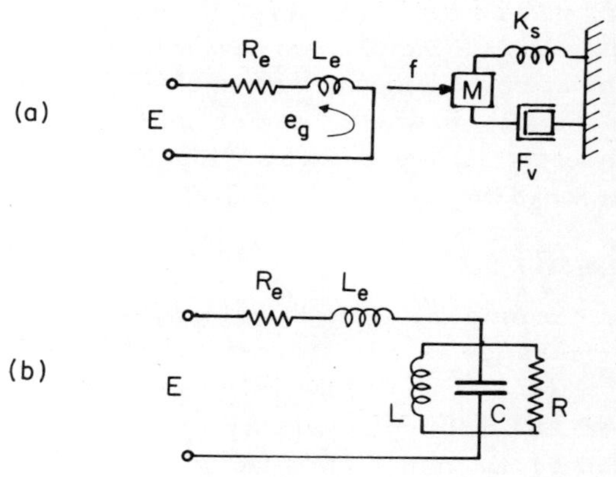

Fig. 2.3.

Note that R_e and L_e are the inherent electrical resistance and inductance of the electrical part, determined from electrical measurements when the device is clamped.

Frequency-response analysis

The electrical impedance measured at the input terminals or

'driving point' impedance can be written,

$$Z_t = Z_e + Z_{\text{mot}}$$

where Z_{mot} is the 'motional impedance' due to the mechanical part and Z_e is the impedance of the electrical part alone.

For the mechanical part we can write:

$$\frac{M}{K_0^2} \cdot \frac{\mathrm{d}e}{\mathrm{d}t} + \frac{F_v}{K_0^2} \cdot e + \frac{K_s}{K_0^2} \int e \cdot \mathrm{d}t = i.$$

Assume that the forcing function is a periodic function of frequency, ω and thus,

$$i = I\epsilon^{j\omega t}$$

Also

$$e = iZ = IZ\epsilon^{j\omega t}$$

$$\frac{\mathrm{d}e}{\mathrm{d}t} = j\omega \cdot IZ\epsilon^{j\omega t}$$

$$\int e \cdot \mathrm{d}t = \frac{1}{j\omega} IZ\epsilon^{j\omega t}.$$

Hence $\dfrac{M}{K_0^2} j\omega \cdot IZ\epsilon^{j\omega t} + \dfrac{F_v}{K_0^2} IZ\epsilon^{j\omega t} + \dfrac{K_s}{K_0^2} \cdot \dfrac{1}{j\omega} IZ\epsilon^{j\omega t} = I\epsilon^{j\omega t}.$

Then, $\dfrac{1}{K_0^2}\left(j\omega M + F_v + \dfrac{K_s}{j\omega}\right) = \dfrac{1}{Z}$ and Z is Z_{mot}.

The term in brackets, sometimes written,

$$\left(F_v + j\omega M - \frac{j}{\omega c_m}\right) = z_m$$

is termed the Mechanical Impedance, (in mechanical units), and $c_m = 1/K_s$, the spring compliance.

Thus the motional impedance, $Z_{\text{mot}} = \dfrac{K_0^2}{z_m}.$

Frequency loci

The characteristics of a transducer may be determined from measurements of its terminal impedance. The value of the impedance will be affected by the conditions influencing the mechanical part. In the case of a moving-coil device the investigation is carried out by supplying an alternating current of various known frequencies to the coil, when the moving element will vibrate in a manner depending on the constants of

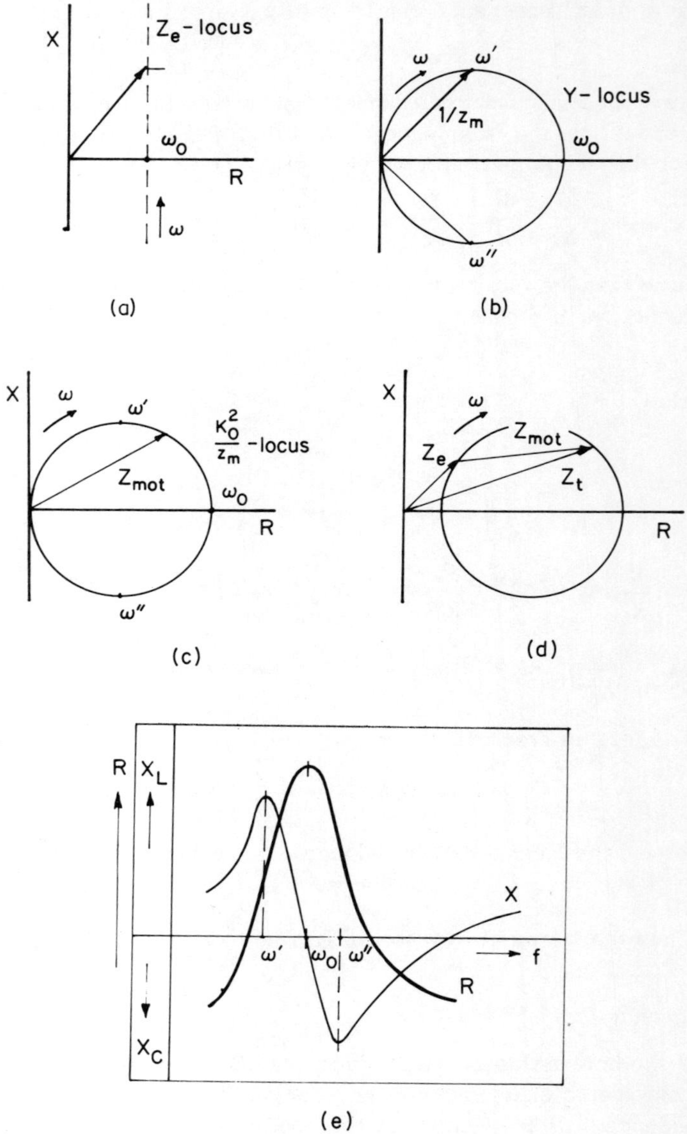

Fig. 2.4.

the system. The transducer is connected as one arm of an impedance-measuring bridge. Sets of driving-point impedance measurements are made and are analysed as follows : Fig. 2.4.

The frequency locus of an impedance

$$Z = R + j\left(\omega L - \frac{1}{\omega C}\right)$$

is a straight line parallel to the j-axis and distant R from it. For admittance $Y = 1/Z$ the locus is a circle of diameter $1/R$, (i.e. the inversion of a straight line). Hence the frequency locus of Z_e for the electromagnetic device considered will be a straight line. This locus can be obtained by clamping the moving element. It will be found to depart somewhat from the ideal locus.

The frequency locus of

$$Z_{\text{mot}} = \frac{K_0^2}{z_m},$$

will be a circle, (i.e. admittance locus modified by the scale factor K_0^2). Hence the locus obtained for the device will be the combined loci, being the locus of

$$Z_t = Z_e + \frac{K_0^2}{Z_m}. \text{ (Fig. 2.4 (d))}.$$

Note that the 'transduction' or 'force' factor, K_0 is a vector and may be expressed, $K_0 = A\epsilon^{-j\beta}$, whence operation upon the mechanical admittance by K_0^2 causes a vector rotation of 2β on the admittance circle diameter. β is generally very small. Again, the motional impedance locus obtained is not a circle exactly. The departures from ideal loci are associated with eddy current and hysteresis effects in electromagnetic devices and with dielectric absorbtion in devices of an electrostatic nature. But sufficient data is obtained from experimental results to permit evaluation of the system constants.

Evaluation of constants

Consider the circular part of the ideal locus for Z_t, i.e. the locus of Z_{mot}. In terms of an equivalent circuit of C, R and L in parallel, Fig. 2.3(b),

$$Y_{\text{mot}} = \frac{1}{R} + j\left(\omega C - \frac{1}{\omega L}\right)$$

$$\therefore \quad Z_{\text{mot}} = \frac{1}{Y_{\text{mot}}} = \frac{R}{1 + jR\left(\omega C - \frac{1}{\omega L}\right)}.$$

Rationalising gives,

$$Z_{\text{mot}} = \frac{R - jR^2\left(\omega C - \dfrac{1}{\omega L}\right)}{1 + R^2\left(\omega C - \dfrac{1}{\omega L}\right)^2}.$$

At resonance, frequency ω_0, the j-component is zero and,

$$\omega_0 C = \frac{1}{\omega_0 L}, \text{ whence } \omega_0^2 = \frac{1}{LC}.$$

Also $Z_{\text{mot}} = R$ at resonance.

When real and reactive parts are equal, the so-called 'quadranta frequencies', ω' and ω'' occur. Since R is positive, algebraically we have,

$$\frac{1}{R} = -\omega' C + \frac{1}{\omega' L} = -\omega' C + \frac{\omega_0^2 C}{\omega'}$$

and
$$\frac{1}{R} = \omega'' C - \frac{1}{\omega'' L} = \omega'' C - \frac{\omega_0^2 C}{\omega''}.$$

By algebraic manipulation,

$$-\omega' + \frac{\omega_0^2}{\omega'} = \omega'' - \frac{\omega_0^2}{\omega''}$$

$$(\omega'' + \omega') = \omega_0^2\left(\frac{1}{\omega'} + \frac{1}{\omega''}\right) = \omega_0^2\left(\frac{\omega'' + \omega'}{\omega'\omega''}\right)$$

$$\underline{\omega'\omega'' = \omega_0^2}.$$

Also,
$$\frac{2}{CR} = (\omega'' - \omega') + \omega_0^2\left(\frac{1}{\omega'} - \frac{1}{\omega''}\right)$$

$$= (\omega'' - \omega') + \omega_0^2\left(\frac{\omega'' - \omega'}{\omega'\omega''}\right)$$

$$\underline{\frac{1}{CR} = (\omega'' - \omega')}.$$

Using the above relations values of R, C and L can be determined from measurements at ω_0, ω' and ω'' taken from the experimental frequency-response locus.

The appropriate values of the mechanical constants, in mechanical units, are determined either by:
(i) one of the mechanical constants is determined by an independant measurement, (e.g. weighing the mass), or
(ii) the Added-mass procedure, viz.,

A small known mass is attached to the moving element. The new resonant frequency, ω_{0A} is determined.

Then $$\omega_{0A}^2 = \frac{1}{L\,(C + \delta C)}.$$

But $$\omega_0^2 = \frac{1}{LC}$$

$$\therefore\ C + \delta C = C\left(\frac{\omega_0^2}{\omega_{0A}^2}\right)$$

and $$\delta C = C\left(\frac{\omega_0^2}{\omega_{0A}^2}\right) - 1\ .$$

Now, $\delta C = \dfrac{\delta M}{K_0^2}$, so that K_0^2 is determined.

Evaluation of the mechanical quantities, in mechanical units follows from the relations:

$$M = K_0^2 C,\ F_v = \frac{K_0^2}{R},\ K_s = \frac{K_0^2}{L},$$

all in m.k.s. units.

N.B. If Z_{mot} had been interpreted as a series R, L, C circuit, a similar form of analysis gives $R/L = (\omega'' - \omega')$. But this means that the mass/inductance analogy has been used, i.e. a node system in analogy with a loop circuit. The analogous equation would be:

$$L\frac{\mathrm{d}i}{\mathrm{d}t} + iR + \frac{1}{C}\int i \cdot \mathrm{d}t = e.$$

(Ref. 4).

REFERENCES

1. KARPLUS, W.J. and SOROKA, W.W. (1959). *Analog Methods*, 2nd ed., McGraw-Hill, New York.
2. SEELY, S. (1962). *Electromechanical Energy Conversion*, McGraw-Hill, New York.
3. MASON, W.P. (1948). *Electromechanical Transducers and Wave Filters*, Van Nostrand, D., New York.
4. HUNT, F.V. (1954). *Electroacoustics*, Harvard Monograph, Wiley, New York.

EXAMPLES

EXAMPLE 2.1.

A mass M is attached to a support by a spring of stiffness K_s and a dashpot attached between the mass and the support has a damping coefficient F_v. (a) Justify the derivation of a mass-capacitance analogue of the system and (b) determine the appropriate scaling factors and obtain the values of the components of the electrical analogue of a system having $M = 1\,\text{kg}$, $K_s = 100\,\text{N/m}$ and $F_v = 1\,\text{N sec/m}$. The time scale chosen gives $t_e = 0{\cdot}001\, t_m$ and the resistor used $= 1000\,\Omega$.

Solution :

(a) The mechanical system has component forces summed at a node and the dynamic equilibrium equation is :

$$M\frac{\mathrm{d}v}{\mathrm{d}t} + F_v \cdot v + K_s\int v \cdot \mathrm{d}t = f.$$

The mass-capacitance analogue is obtained by summing current components at a node, viz.

$$C\frac{\mathrm{d}e}{\mathrm{d}t} + \frac{1}{R}\cdot e + \frac{1}{L}\int e \cdot \mathrm{d}t = i.$$

Hence analogous quantities are as similarly placed in the two algebraic equations.

(b) The primary mechanical measurements M, l, t_m are related to analogous electrical quantities by arbitary scale factors, thus :

$$C = k_1 M; \int e \cdot \mathrm{d}t = k_2 l; \; t_e = k_3 t_m .$$

Now $1/R \propto F_v$ say $1/R = k_4 F_v$ and F_v has dimensions of force/velocity $= [Mlt_m^{-2} \times t_m/l] = [M/t_m]$.

To convert F_v to electrical terms gives $k_1 M/k_3 t_m$. Hence

$$k_4 = \frac{k_1}{k_3}.$$

We are given $R = 1000\,\Omega$ and $k_3 = 10^{-3}$

$$\therefore \quad \frac{1}{R} = 10^{-3} = k_4 F_v = k_4 (1)$$

$$= k_1/k_3.$$

Thus $$k_1 = k_3 \times 10^{-3} = 10^{-6}$$

$$\therefore \quad C = k_1 M = 10^{-6}\,\text{F} = 1\,\mu\text{F}.$$

Now say
$$L = k_5 \frac{1}{K_s}$$

and since $1/K_s$ = distance/force = $[l/Mlt_m^{-2}] = [t_m^2/M]$

$\therefore$ Scale factor $k_5 = k_3^2/k_1 = 10^{-6}/10^{-6}$

$\therefore$ $L = 1/100 = 0{\cdot}01\,\text{H} = 10\,\text{mH}$.

EXAMPLE 2.2.

Explain how the 'quadrantal frequencies' are employed in the derivation of the constants for the equivalent electrical circuit of an electro-mechanical energy transducer. In a test the measured values of these frequencies were given as 187 c/s and 267 c/s. The ohmic resistance due to the mechanical part was 10 Ω. Determine (a) the equivalent electrical circuit of the mechanical part as a mass-capacitance analogue, (b) the constants in mechanical units for the mechanical part if an added mass of 1 g changes the resonant frequency to $(1/1{\cdot}2)^{\frac{1}{2}}$ times the original value.

Theory part: Reference to p. 44 and Fig. 2.4 (c), it is shown that

$$1/CR = (\omega'' - \omega') \text{ and } \omega''\omega' = \omega_0^2 .$$

Also
$$\omega_0^2 = 1/LC.$$

For part (b), with known added mass, δM,

$$\delta C = C(\omega_0^2/\omega_{0A}^2 - 1) = \frac{\delta M}{K_0^2}$$

from which K_0^2 is determined, and the mechanical components.

Numerical part, (a) $f' = 187$ c/s $\therefore$ $\omega' = 1180$

$f'' = 267$ c/s $\therefore$ $\omega'' = 1680$

Substituting, $1/CR = (\omega'' - \omega') = 500$

R is given $= 10\,\Omega$ $\therefore$ $C = 200\,\mu\text{F}$

$\omega_0^2 = \omega''\omega' = 198 \cdot 10^4 = 1/LC$

$\therefore$ $L = 2{\cdot}5\,\text{mH}$.

The equivalent electrical circuit of the mechanical part as a M–C analogue has R, L and C arranged in parallel and having the above values.

Numerical part (b). From the above equation,

$$\delta C = 200\,(0{\cdot}2) = 40\,\mu\text{F}$$

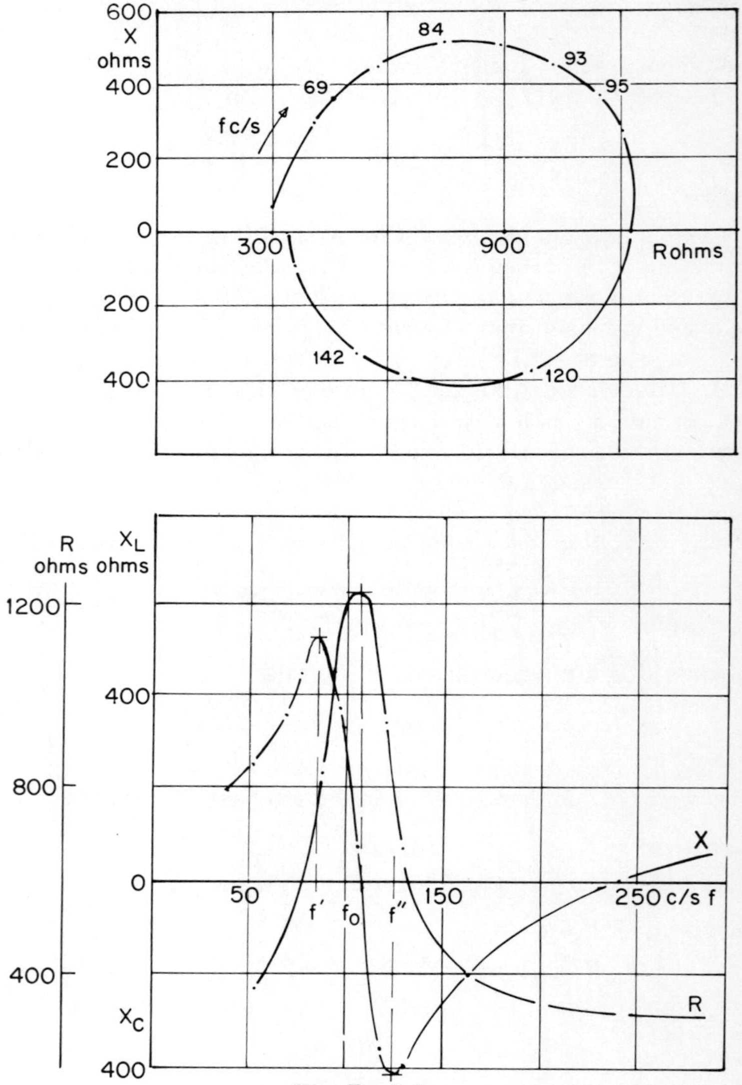

Fig. Ex. 2.3.

and this represents 1 g.

$$\therefore \quad M, \text{ represented by } 200\,\mu\text{F} = 5\,\text{g}.$$

Now $C = M/K_0^2$. $\therefore \quad K_0^2 = 5 \times 10^{-3}/200 \times 10^{-6} = 25$ m.k.s. units.

$$\text{Then } F_v = K_0^2/R = 25/10 = 2{\cdot}5\text{N sec/m}.$$
$$K_s = K_0^2/L = 25/2{\cdot}5 \times 10^{-3} = 10\,000\,\text{N/m}.$$

EXAMPLE 2.3.

The experimental values obtained from impedance measurements carried out on an electromechanical vibrator unit are shown in Fig. Ex. 2.3. The measurements were made using a variable-frequency oscillator, a Maxwell bridge for inductive-reactance balance and this was changed to a de Sauty bridge when the reactance became capacitative. Determine the values of the mechanical constants of the vibrator unit given that the mass of the moving element is 10 g.

Solution:

From the motional impedance circle, diameter $= 880\Omega = R$.

Note that greater accuracy is obtained in determining ω_0, ω' and ω'' if the results are plotted on a frequency axis, Fig. Ex. 2.3.

$$\omega'' = 2\pi \times 125 = 785 \text{ rad/sec}$$
$$\omega_0 = 2\pi \times 107 = 672 \text{ rad/sec}$$
$$\omega' = 2\pi \times 85 = 534 \text{ rad/sec}$$

$$\therefore \quad C = 1/R(\omega'' - \omega') = 1/(880 \times 251) = 4{\cdot}55\,\mu\text{F}.$$

This is a M–C analogy, $C = M/K_0^2$ and $M = 10$ g, (given).

In m.k.s. units,

$$K_0^2 = M/C = 10 \times 10^{-3}/4{\cdot}55 \times 10^{-6} = 2200$$

$$\therefore \quad \text{Viscous friction, } F_v = K_0^2/R = 2200/880 = \underline{2{\cdot}5\text{N sec/m}}$$

$$\omega_0^2 = 1/LC \;\therefore\; L = 1/(672)^2 \times 4{\cdot}55 \times 10^{-6} = 0{\cdot}485\,\text{H}$$

$$\therefore \quad \text{Spring constant, } K_s = K_0^2/L = 2200/0{\cdot}485 = \underline{4540\,\text{N/m}}.$$

3 Energy conversion and inductance

The study of electromagnetic transducers is essentially the study of the interaction of electric circuits and magnetic fields and the associated energies. Energy conversion occurs due to forces arising between the fixed and the moving parts of the transducer. The concepts of self and mutual inductance embrace both circuit and field relations and consequently can provide a compact form of analysis which proves to be useful both in the study of systems and in the design of devices or machines.

First we examine the development of force and energy in a transducer, based upon the concept of 'field energy' and the principle of 'virtual displacement'. To relate force and energy requires that the point of application of the force should move and thereby do work. The method of virtual displacement assumes a differentially small displacement, dx, during which the force is assumed to remain constant. In the limit, as d$x \rightarrow 0$, the assumption is valid and the method permits an energy balance to be stated.

Field energy

The relation between magnetomotive force and flux for a magnetic system having a ferromagnetic core is non-linear. In terms of magnetising force H, and flux density B, this relation is expressed:

$$B = \mu_0 \mu_r H$$

where $\mu_0 = 4\pi \cdot 10^{-7}$ and μ_r, the relative permeability, is a parameter.

Because of hysteresis, (Greek: lagging), and eddy currents, the relation is not single valued. Magnetisation occurs due to the re-orientation of domain magnetic axes. Energy expended in domain changes may be stored, resulting in the phenomenon of 'residual magnetism'.

Under cyclic conditions of magnetisation, energy is expended and the area of the 'hysteresis loop' is a measure of the energy expended per cycle, viz;

$$\begin{aligned}\text{Energy supplied} &= ei \cdot \mathrm{d}t \\ &= N \cdot \frac{\mathrm{d}\phi}{\mathrm{d}t} \cdot i \cdot \mathrm{d}t \\ &= Hl \cdot \mathrm{d}(BA) \\ &= \int H \cdot \mathrm{d}B. \text{ (volume) joules.}\end{aligned}$$

Empirically it can be shown for the 'static' hysteresis loop,

$$\text{Hysteresis loss} = k_h \cdot B^x f \text{W/m}^3$$

where k_h is a constant and x is the Steinmetz index. If the loop is obtained 'dynamically', eddy current losses are added and the loop area is larger.

Eddy currents are induced in the iron mass due to rapid flux changes. The energy due to these currents causes heating and is not recoverable. Accurate calculation of eddy-current losses is difficult, but an elementary appreciation can be obtained as follows:

$$\text{Induced eddy voltage} \propto Bf$$

$$\begin{aligned}\text{Eddy } i^2 r \text{ loss} &\propto \frac{B^2 f^2}{r} \\ &= k_e B^2 f^2 \text{ (volume) watts.}\end{aligned}$$

Magnetic cores subject to rapidly changing magnetic fluxes may be made from low-loss iron alloy and are commonly laminated to minimise eddy current circulation and losses.

During the magnetic build-up process, the energy, W_{fld}, absorbed by the field is given by the area between the curve and the flux axis, (area Oac, Fig. 3.1(b)). The area under the curve is termed the Co-energy, (W'_{fld}). When the m.m.f. is reduced to zero the energy returned to the circuit is given by the area (abc).

Hence the nett energy stored in the field is (Oab).

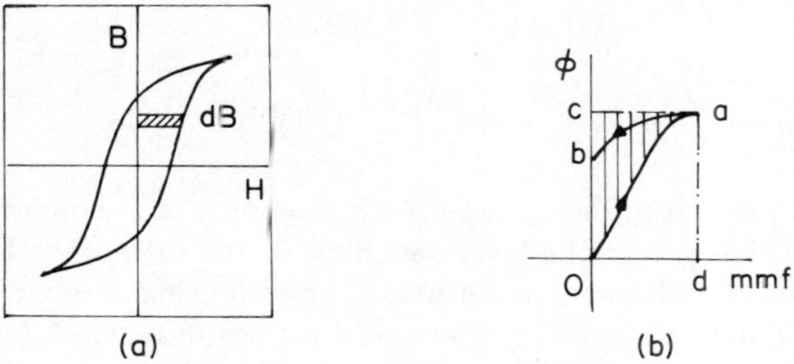

Fig. 3.1.

Now consider an elementary relay or contactor mechanism, Fig. 3.2(a). When the moving armature operates a magnetic force, f, acts to close the magnetic circuit. The area between the magnetisation curves corresponding to open and closed conditions represents the change in stored field energy, Fig. 3.2(b).

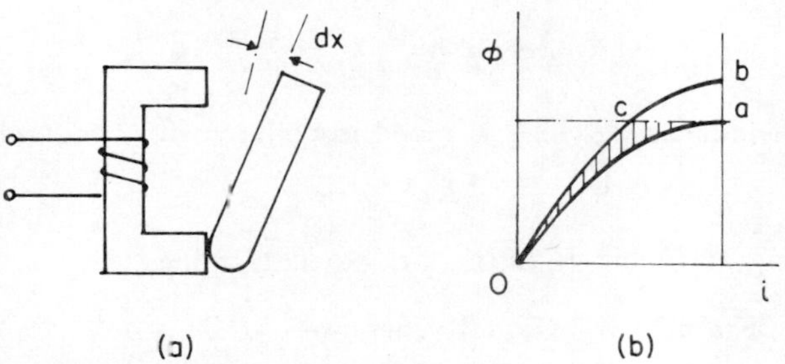

Fig. 3.2.

For a differential movement, dx,

$$\text{Mechanical work done} = f \cdot \mathrm{d}x$$

$$\text{Change in field energy} = \text{OabcO, (at constant } i\text{)},$$

$$\text{or} = \text{OacO, (at constant } \phi\text{)}.$$

An exact assessment of the changes in energy is complicated since, for example, during the transient period an e.m.f. will be induced in the coil, for changing flux. Also, at constant flux conditions the change in stored energy is $\mathrm{d}W_{fld}$, but at constant current conditions the change is seen to be an increase in co-energy, dW'_{fld}. These conditions may be expressed explicitly in functional notation:

$$f = -\frac{\partial W_{fld}}{\partial x}\ (\phi, x)$$

$$\text{and}\quad f = +\frac{\partial W'_{fld}}{\partial x}\ (i, x).$$

The sign preceding the change in field energy will depend on the chosen variables. In the limit, as $dx \to 0$, the difference between constant-flux and constant-current change become second-order effects. These mathematical constraints are discussed fully in Ref. 1.

In a magnetic circuit with an airgap, most of the m.m.f. is expended on the airgap and most of the energy is stored in the airgap. Neglecting core losses and non-linearity,

$$W_{fld} = \tfrac{1}{2}\Phi\mathcal{F} = \tfrac{1}{2}\lambda i$$

$$\text{But reluctance} = \frac{\text{m.m.f.}}{\text{flux}},\ \text{or}\ \mathcal{R} = \frac{\mathcal{F}}{\Phi}$$

$$\therefore\ W_{fld} = \tfrac{1}{2}\mathcal{R}\Phi^2 = \tfrac{1}{2}\frac{1}{\mathcal{R}}\mathcal{F}^2$$

Now self-inductance, $L = \lambda/i$ and is constant for constant R

$$\therefore\ W_{fld} = \tfrac{1}{2}Li^2$$

Working in B and H units, i.e. per unit volume,

$$\text{Field Energy Density, } w_{fld} = \tfrac{1}{2}HB = \tfrac{1}{2}\frac{B^2}{\mu_0}.$$

The linear-motion device

Consider a simple solenoid-type or a unipivot relay mechanism, (e.g. Fig. 3.2(a)), where the motion may be assumed to be linear. To minimise mathematical complexity, leakage flux paths and non-linearities are neglected. A further simplification is achieved by assuming that all the field energy is stored in the airgap, i.e. $\mu_{iron} \gg \mu_{air}$.

Let x = length of airgap,
A = cross-section area of airgap.

Then, Volume of airgap $= Ax$.

$$\text{Stored field energy} = \tfrac{1}{2}\frac{B^2}{\mu_0} \cdot Ax.$$

If the armature or moving part is moved to a new gap $= (x + \mathrm{d}x)$.

$$\text{New stored energy} = \tfrac{1}{2}\frac{B^2}{\mu_0} \cdot A(x + \mathrm{d}x).$$

$$\text{Change in stored energy} = \tfrac{1}{2}\frac{B^2}{\mu_0} \cdot A \cdot \mathrm{d}x.$$

$$\text{But work done} = F \cdot \mathrm{d}x$$

$$\therefore \quad F = \tfrac{1}{2}\frac{B^2}{\mu_0} A$$

In terms of current, since $B = \mu_0 H = \mu_0 \dfrac{Ni}{x}$

$$\therefore \quad F = \tfrac{1}{2}\frac{\mu_0 N^2 i^2 A}{x^2}.$$

Note: The force increases as x decreases. Hence the iron is drawn into the region of greater energy density. This may be interpreted:

(a) that there is a magnetic pressure on a surface between two regions of different permeabilities in a magnetic field, and:

$$\text{Magnetic pressure} = \text{force/unit area,}$$

or

$$F/A = B^2/2\mu_0 = \tfrac{1}{2}HB.$$

(b) since energy density may be written $= \int H \cdot \mathrm{d}B$, the force will depend on the divergence of the field.

A general equation of energy balance may be stated:

Mech. input + Elect. input = Mech. storage + Field storage + Losses

$$F \cdot \mathrm{d}x + ei \cdot \mathrm{d}t = 0 + \frac{B^2}{2\mu_0} A \cdot \mathrm{d}x + i^2R \cdot \mathrm{d}t.$$

With the limitations of virtual displacement, mech. storage arising from motion is zero, whilst elect. input is dissipated in ohmic losses, there being no flux change.

N.B. if F is applied, mech. input and field storage are $+ve$

if F is derived, mech. input and field storage are $-ve$.

Reluctance force:

The role of variation of reluctance in the development of force is important. Where $\mathcal{R}$ and Φ are variables,

$$\mathrm{d}W_{fld} = \tfrac{1}{2}\,\Phi^2 \cdot \mathrm{d}\mathcal{R} + \mathcal{R}\Phi \cdot \mathrm{d}\Phi$$
$$= \tfrac{1}{2}\,\Phi^2 \cdot \mathrm{d}\mathcal{R} + \mathcal{F} \cdot \mathrm{d}\Phi.$$

Also, elect. input $= ei \cdot \mathrm{d}t = N \cdot \dfrac{\mathrm{d}\Phi}{\mathrm{d}t} \cdot i \cdot \mathrm{d}t$

$$= Ni \cdot \mathrm{d}\Phi = \mathcal{F} \cdot \mathrm{d}\Phi.$$

On energy balance:

$$\mathrm{d}W_{\text{mech}} + \mathrm{d}W_{\text{elect}} = \mathrm{d}W_{fld}$$
$$F \cdot \mathrm{d}x + \mathcal{F} \cdot \mathrm{d}\Phi = \tfrac{1}{2}\,\Phi^2 \cdot \mathrm{d}\mathcal{R} + \mathcal{F}\mathrm{d}\Phi$$
$$F \cdot \mathrm{d}x = \tfrac{1}{2}\,\Phi^2 \cdot \mathrm{d}\mathcal{R}$$
$$\therefore\ F = \tfrac{1}{2}\,\Phi^2 \cdot \frac{\mathrm{d}\mathcal{R}}{\mathrm{d}x}$$

The rotational equivalent is: $\textcircled{T} = \tfrac{1}{2}\,\Phi^2 \cdot \dfrac{\mathrm{d}\mathcal{R}}{\mathrm{d}\theta}.$

Also, since $W_{fld} = \tfrac{1}{2}\,Li^2, \quad \therefore\ \textcircled{T} = \tfrac{1}{2}\,i^2 \cdot \dfrac{\mathrm{d}L}{\mathrm{d}\theta}.$

Note: in motoring concepts, F and $\textcircled{T}$ are derived, and in the above notation would be $-ve$.

The rotational device

The principles of energy density and virtual displacement are now applied to a singly-excited rotational device. An airgap occurs at each end of the rotor, Fig. 3.3. Then in general:

$$\text{Volume of airgaps} = 2\,(r\theta x l)$$
$$\text{Stored field energy} = (2r\theta x l)\,\frac{B^2}{2\mu_0} = \frac{x\Phi^2}{\mu_0 r\theta l}.$$

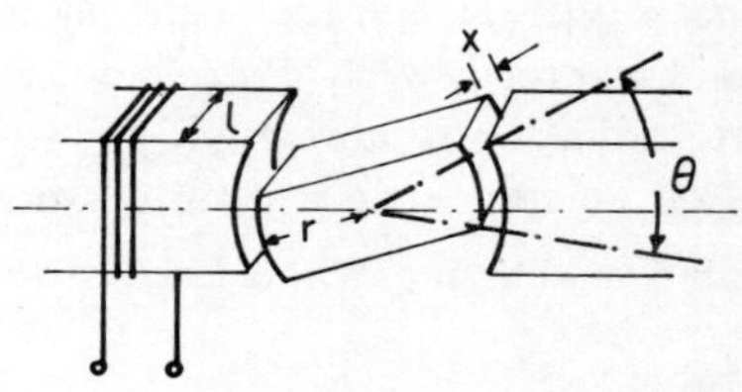

Fig. 3.3.

(*i*) For constant Φ:

Note that stored energy decreases as θ increases.

$$\frac{\mathrm{d}}{\mathrm{d}\theta}(W_{fld})\cdot\mathrm{d}\theta = -\frac{x\Phi^2}{\mu_0 r\theta^2 l}\cdot\mathrm{d}\theta = Ⓣ\cdot\mathrm{d}\theta$$

$$\therefore\quad Ⓣ = -\frac{x\Phi^2}{\mu_0 r\theta^2 l}.$$

The $-ve$ sign indicates that the device acts as a motor.

$$\text{Thus, Output torque} - Ⓣ = \frac{x\Phi^2}{\mu_0 r\theta^2 l}.$$

$$\text{In terms of } B,\quad -Ⓣ = \frac{xrl}{\mu_0}\cdot B^2.$$

$$\text{In terms of } Ni,\text{ since } B = \mu_0 H = \mu_0\frac{Ni}{2x}$$

$$-Ⓣ = \frac{\mu_0 r l}{4x}\cdot N^2 i^2.$$

(*ii*) For constant current:

Since $B = \mu_0\cdot\dfrac{Ni}{2x}$, we can write,

$$\text{Stored field energy, } W_{fld} = \mu_0\frac{rl}{4x}\cdot N^2 i^2\theta.$$

Then for variable θ, constant i,

$$\frac{\mathrm{d}}{\mathrm{d}\theta}(W_{fld})\cdot\mathrm{d}\theta = \mu_0\frac{rl}{4x}\cdot N^2 i^2\cdot\mathrm{d}\theta.$$

With constant current as θ changes the flux changes, causing an induced e.m.f., e opposing i. To maintain constant i there must be a supply of energy. Then electrical input in time $\mathrm{d}t$ for a change $\mathrm{d}\theta$,

$$= ei\cdot\mathrm{d}t = N\frac{\mathrm{d}\Phi}{\mathrm{d}t}\cdot i\cdot\mathrm{d}t = Ni\cdot\mathrm{d}\Phi.$$

But $$\Phi = rl\theta B = \frac{\mu_0 r l}{2x}\cdot Ni\cdot\theta.$$

$$\therefore\quad \text{Elect. input} = \frac{\mu_0 r l}{2x}\cdot N^2 i^2\cdot\mathrm{d}\theta.$$

Neglecting heat losses and mechanical storage,

$$Ⓣ \cdot d\theta + \frac{\mu_0 r l}{2x} \cdot N^2 i^2 \cdot d\theta = \frac{\mu_0 r l}{4x} \cdot N^2 i^2 \cdot d\theta.$$

Hence, as before,
$$-Ⓣ = \frac{\mu_0 r l}{4x} \cdot N^2 i^2.$$

Note: at constant current,

$\frac{1}{2}$ elect. energy supplied → mech. energy

$\frac{1}{2}$ elect. energy supplied → field storage.

Torque and inductance

Electromagnetic energy-conversion processes may be interpreted in terms of inductance, in forms which are simpler and of more general application than those which are defined in terms of physical or design details

(a) Consider a singly-excited system.

In the elementary device shown in Fig. 3.3., when the coil is energized a torque acts on the ferromagnetic rotor tending to align the rotor axis with the field axis. This action was interpreted in terms of airgap dimensions, or in terms of $d\mathcal{R}/d\theta$, that is a reluctance torque.

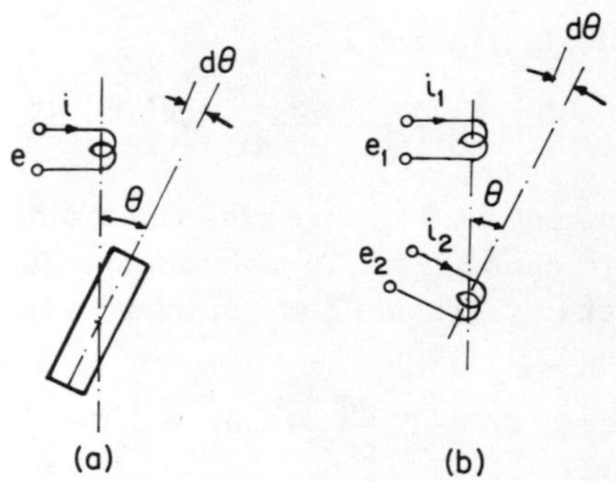

Fig. 3.4.

The device may be symbolised more simply as in Fig. 3.4 (a), and the exciting coil has resistance R and self-inductance L.

Then we have, input voltage, $e = iR + L\dfrac{di}{dt}$ if L is constant,

$$= iR + \frac{d(Li)}{dt} \quad \text{if } L \text{ is variable.}$$

$$\text{Electrical input energy} = ei \cdot dt = \left[iR + \frac{d(Li)}{dt}\right] i \cdot dt$$

$$= i^2R \cdot dt + Li \cdot di + i^2 \cdot dL.$$

Now, energy stored in a magnetic field $= \frac{1}{2}Li^2$.

Change of stored field energy in time dt,

$$= \frac{d(\frac{1}{2}Li^2)}{dt} \cdot dt$$

$$= Li \cdot di + \tfrac{1}{2} i^2 \cdot dL.$$

Using the energy balance equation, and allowing that the mechanical energy developed supplies the energy of mechanical inertia, (if included),

$$\therefore \text{ Mech. energy developed} = \tfrac{1}{2} i^2 \cdot dL$$

$$= \textcircled{T} \cdot d\theta$$

$$\therefore \quad \textcircled{T} = \tfrac{1}{2} i^2 \cdot \frac{dL}{d\theta}.$$

The linear equivalent is: $\quad F = \frac{1}{2} i^2 \cdot \dfrac{dL}{dx}.$

Note: (*i*) energy due to differential current changes, (di), does not affect the mechanical forces.

(*ii*) at constant i, $\frac{1}{2}$ energy supplied – mechanical energy

$\frac{1}{2}$ energy supplied – field storage.

(*iii*) Taking the simple linear case,

$$L = \frac{N\Phi}{i} \quad \text{and} \quad \Phi = BA = \frac{\mu_0 NiA}{x}$$

$$\therefore \quad L = \frac{\mu_0 N^2 A}{x} \quad \text{and} \quad \frac{dL}{dx} = -\frac{\mu_0 N^2 A}{x^2}$$

Then $\quad F = \frac{1}{2} i^2 \cdot dL/dx = -\dfrac{\mu N^2 i^2 A}{2x^2}$ as before.

(b) Doubly or Multiple-excited systems.

Many electromagnetic transducers have more than one winding, usually arranged with one winding or a group of windings on the fixed part and another winding or a group of windings on the other or moving part. Most rotating machines for electromechanical energy conversion are of this type.

Self and mutual inductances are involved and these are functions of the angular position of the rotor. Differential changes in energies due to a virtual displacement, $d\theta$, of the rotor are examined for the doubly-excited case, Fig. 3.4 (b).

Let L_1, L_2 = self-inductances of the individual windings,

M = mutual inductance between the two windings.

Electrical input, $dW_{elect} = (e_1 i_1 + e_2 i_2)\,dt,$

where
$$e_1 = i_1 R_1 + \frac{d}{dt}(L_1 i_1) + \frac{d}{dt}(Mi_2)$$

$$e_2 = i_2 R_2 + \frac{d}{dt}(L_2 i_2) + \frac{d}{dt}(Mi_1)$$

Then,
$$dW_{elect} = \left[i_1 R_1 + \frac{d}{dt}(L_1 i_1) + \frac{d}{dt}(Mi_2)\right] i_1 \cdot dt + \left[i_2 R_2 + \frac{d}{dt}(L_2 i_2) + \frac{d}{dt}(Mi_1)\right] i_2 \cdot dt$$

$$= (i_1^2 R_1 + i_2^2 R_2)\,dt + i_1^2 \cdot dL_1 + i_2^2 \cdot dL_2 + 2 i_1 i_2 \cdot dM + (L_1 i_1 + Mi_2)\,di_1 + (L_2 i_2 + Mi_1)\,di_2$$

Field storage energy, $W_{fld} = \frac{1}{2} L_1 i_1^2 + \frac{1}{2} L_2 i_2^2 + Mi_1 i_2.$

Change in $W_{fld} = dW_{fld} = \frac{1}{2} i_1^2 \cdot dL_1 + \frac{1}{2} i_2^2 \cdot dL_2 + i_1 i_2 \cdot dM + (L_1 i_1 + Mi_2)\,di_1 + (L_2 i_2 + Mi_1)\,di_2.$

Applying the energy balance equation :

Mechanical energy developed $= \frac{1}{2} i_1^2\, dL_1 + \frac{1}{2} i_2^2\, dL_2 + i_1 i_2\, dM.$

$$= Ⓣ\, d\theta.$$

Hence
$$Ⓣ = \left[\frac{1}{2} i_1^2 \frac{dL_1}{d\theta} + \frac{1}{2} i_2^2 \frac{dL_2}{d\theta}\right] + i_1 i_2 \frac{dM}{d\theta}$$

= Reluctance torque, $Ⓣ_r$ + Excitation torque, $Ⓣ_e$.

Also, $Ⓣ = J \dfrac{d^2\theta}{dt^2} + Ⓣ_{mech}.$

The corresponding voltage equations for the coils are:

$$\begin{aligned} e_1 &= \frac{\mathrm{d}}{\mathrm{d}t}(L_1 i_1 \pm M i_2) \\ &= L_1 \frac{\mathrm{d}i_1}{\mathrm{d}t} + i_1 \frac{\mathrm{d}L_1}{\mathrm{d}t} \pm M \frac{\mathrm{d}i_2}{\mathrm{d}t} \pm i_2 \frac{\mathrm{d}M}{\mathrm{d}t} \\ &= \left(L_1 \frac{\mathrm{d}i_1}{\mathrm{d}t} \pm M \frac{\mathrm{d}i_2}{\mathrm{d}t}\right) + \left(i_1 \frac{\mathrm{d}L_1}{\mathrm{d}t} \pm i_2 \frac{\mathrm{d}M}{\mathrm{d}t}\right) \\ &= \left(L_1 \frac{\mathrm{d}i_1}{\mathrm{d}t} \pm M \frac{\mathrm{d}i_2}{\mathrm{d}t}\right) + \left(i_1 \frac{\mathrm{d}L_1}{\mathrm{d}\theta} \pm i_2 \frac{\mathrm{d}M}{\mathrm{d}\theta}\right)\frac{\mathrm{d}\theta}{\mathrm{d}t} \\ &= \text{Transformer volts} + \text{Rotational volts.} \end{aligned}$$

Similarly:

$$e_2 = \left(L_2 \frac{\mathrm{d}i_2}{\mathrm{d}t} \pm M \frac{\mathrm{d}i_1}{\mathrm{d}t}\right) + \left(i_2 \frac{\mathrm{d}L_2}{\mathrm{d}\theta} \pm i_1 \frac{\mathrm{d}M}{\mathrm{d}\theta}\right)\frac{\mathrm{d}\theta}{\mathrm{d}t}.$$

Generalised form

The results may be summarised for general application and expressed in a compact notation as follows:

Working from first principles, to emphasise the assumptions made, consider one coil coupled to several others, with random geometry, Fig. 3.5.

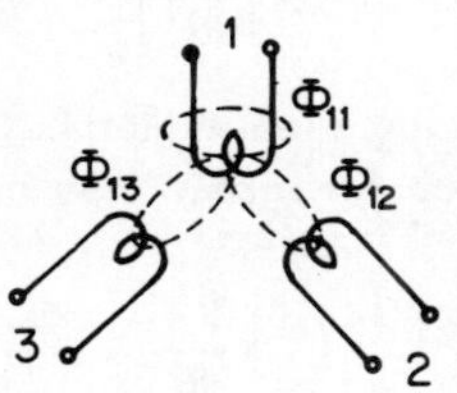

Fig. 3.5.

$$\begin{aligned} e_1 &= i_1 r_1 + \frac{\mathrm{d}}{\mathrm{d}t}(\lambda_{11} + \lambda_{12} + \lambda_{13} + \ldots) \\ &= i_1 r_1 + \frac{\mathrm{d}}{\mathrm{d}t}(N_1 \Phi_{11} + N_1 \Phi_{12} + N_1 \Phi_{13} + \ldots) \end{aligned}$$

where λ = flux linkages and assuming the fluxes link with all N_1 effective turns.

With the further assumption of a linear system,

$$\frac{\mathrm{d}\Phi}{\mathrm{d}i} = \text{constant and } \Phi_{11} = \frac{L_{11} i_1}{N_1},\ \Phi_{12} = \frac{L_{12} i_2}{N_1} \text{ etc.,}$$

and M has been replaced by L_{12}, L_{13} etc., as appropriate, then,

$$e_1 = i_1 r_1 + \frac{d}{dt}(L_{11} i_1 + L_{12} i_2 + L_{13} i_3 + \ldots).$$

Hence, as demonstrated in the doubly-excited case above, and using

$$p = \frac{d}{dt} \text{ and } \omega = \frac{d\theta}{dt}$$

$$e_1 = i_1 r_1 + [L_{11} p i_1 + L_{12} p i_2 + \ldots] + \left[\omega\left(\frac{dL_{11}}{d\theta}\right) i_1 + \omega\left(\frac{dL_{12}}{d\theta}\right) i_2 + \ldots\right].$$

Similar expressions are obtained for e_2, e_3, etc. The equations can be expressed in matrix notation and for the two coil case would be:

$$\begin{vmatrix} e_1 \\ e_2 \end{vmatrix} = \begin{vmatrix} r_1 + L_{11} p + \omega\left(\frac{dL_{11}}{d\theta}\right) & L_{12} p + \omega\left(\frac{dL_{12}}{d\theta}\right) \\ L_{21} p + \omega\left(\frac{dL_{21}}{d\theta}\right) & r_2 + L_{22} p + \omega\left(\frac{dL_{22}}{d\theta}\right) \end{vmatrix} \times \begin{vmatrix} i_1 \\ i_2 \end{vmatrix}$$

Thus, in general,

$$|e| = \left\{ [r] + [L] p + \omega \left[\frac{dL}{d\theta}\right] \right\} [i] \text{ or } [e] = [Z]\,[i]$$

where $[Z]$ is termed the Impedance Matrix.

The general expression for torque may be stated in matrix form:

$$\text{Ⓣ} = \tfrac{1}{2} \begin{vmatrix} i_1 & i_2 \end{vmatrix} \times \begin{vmatrix} \frac{dL_{11}}{d\theta} & \frac{dL_{12}}{d\theta} \\ \frac{dL_{21}}{d\theta} & \frac{dL_{22}}{d\theta} \end{vmatrix} \times \begin{vmatrix} i_1 \\ i_2 \end{vmatrix}$$

or, more briefly,

$$\text{Ⓣ} = \tfrac{1}{2} [i_t] \left[\frac{dL}{d\theta}\right] [i]$$

where $[i_t]$ is termed the 'transpose' of the current matrix.

These equations are important in that they describe the fundamental dynamical behaviour of electromechanical devices. But they are products of variables and derivatives and are non-linear differential equations. However, by the choice of acceptable approximations, they permit the derivation of much important information, as shown below. The generalised forms are further developed in Chapter 8.

Fundamental torque conditions in machines

Tbe basic torque equation for a doubly-excited device, in general terms of inductance, has been shown thus:

$$Ⓣ = \tfrac{1}{2} i_1^2 \frac{dL_1}{d\theta} + \tfrac{1}{2} i_2^2 \frac{dL_2}{d\theta} + i_1 i_2 \frac{dM}{d\theta}.$$

Various possibilities arise depending on the nature of the variables i, L and M, and some examples are considered.

1. The singly-excited reluctance motor.

Being singly-excited, the L_2 and M terms disappear and as before, p. 59.

$$Ⓣ = \tfrac{1}{2} i^2 \frac{dL_1}{d\theta}.$$

For the elementary two-pole case, we may assume that reluctance and inductance vary sinusoidally with rotor displacement, θ.
Let $L_1 = L_0 + L \cos 2\theta$, whence

$$\frac{dL_1}{d\theta} = -2L \sin 2\theta$$

and $i = I_m \cos \omega t$.

Then $$Ⓣ = -LI_m^2 \cos^2 \omega t \cdot \sin 2\theta.$$

Expanding this expression,

$$Ⓣ = -\frac{LI_m^2}{2}(\sin 2\theta + \sin 2\theta \cos 2\omega t)$$

$$= -\frac{LI_m^2}{2}[\sin 2\theta + \tfrac{1}{2}\sin(2\theta + 2\omega t) + \tfrac{1}{2}\sin(2\theta - 2\omega t)].$$

Under running conditions θ will be a function of time and the average of the time-dependant terms will be zero, giving zero nett torque.
But if $\theta = (\omega t - \delta)$ and thus $d\theta/dt = \omega$, the running speed is said to be 'synchronous' with the supply frequency. Then:

$$Ⓣ = -\frac{LI_m^2}{2}[\sin(2\omega t - 2\delta) + \tfrac{1}{2}\sin(4\omega t - 2\delta) + \tfrac{1}{2}\sin(-2\delta)]$$

and $$Ⓣ_{av} = +\tfrac{1}{4} L I_m^2 \sin 2\delta.$$

Note: This 'reluctance' torque is $\propto \sin 2\delta$, where δ, the fixed displacement from synchronous or 'in line' position, is termed the Load or Torque Angle. Hence, where the airgap is non-uniform, as in a machine with 'salient poles', the machine runs at synchronous speed due to reluctance torque. But the machine must be run up to synchronous speed before it locks-in, and develops torque.

2. The synchronous machine.

Assuming uniform air-gap, L_1 and L_2 become constants and,

$$Ⓣ = i_1 i_2 \frac{dM}{d\theta}.$$

With double excitation, one winding is supplied steady d.c., the other has alternating current.

Let $i_1 = I_1$

$i_2 = I_2 \cos \omega t$ where I_2 = max value (a.c.).

Now, assuming sinusoidal variation of M with displacement,

$$M = M_{max} \cos \theta$$

$$\frac{dM}{d\theta} = -M_{max} \sin \theta.$$

Substituting in the expression for Ⓣ:

$$Ⓣ = I_1 I_2 \cos \omega t \,(-M_{max} \sin \theta)$$

$$= -\tfrac{1}{2} I_1 I_2 M_{max} (\sin \overline{\theta + \omega t} + \sin \overline{\theta - \omega t}).$$

At synchronous speed $d\theta/dt = \omega$, and with $\theta = (\omega t - \delta)$.

Hence $Ⓣ = -\tfrac{1}{2} I_1 I_2 M_{max} [\sin(2\omega t - \delta) + \sin(-\delta)]$.

Since the average of time-dependant terms is zero,

Nett average Torque, $Ⓣ_{av} = \tfrac{1}{2} I_1 I_2 M_{max} \sin \delta.$

Note: (a) This is an excitation torque and is $\alpha \sin \delta$.

(b) The machine is not self-starting but must be run up to synchronous speed.

(c) A machine with salient poles and double excitation

develops both reluctance and excitation torques when running at synchronous speed.

3. The induction motor.

With uniform air-gap and double excitation,

$$\text{Ⓣ} = i_1 i_2 \frac{dM}{d\theta}.$$

In this case, with sinusoidal a.c. in both windings:

Let $i_1 = I_1 \cos\omega_1 t$

$i_2 = I_2 \cos\omega_2 t$ where I_1 and I_2 are max values (a.c.)

and $M = M_{max} \sin\theta$ where $\theta = \omega_3 t$.

$\omega_1, \omega_2, \omega_3$ are all different.

Then $\text{Ⓣ} = (I_1 \cos\omega_1 t)(I_2 \cos\omega_2 t)(M_{max} \cos\omega_3 t)$.

Various combinations of the time-dependant terms are possible, and a condition is sought whereby a term will emerge which is not time-dependant, i.e. a nett average torque will be developed. For example, combining the ω_2 and ω_3 terms,

$$\text{Ⓣ} = \tfrac{1}{2} I_1 I_2 M_{max} \cos\omega_1 t[\cos(\overline{\omega_2 + \omega_3}\, t) + \cos(\overline{\omega_2 - \omega_3}\, t)].$$

Then if $\omega_1 = (\omega_2 + \omega_3)$

$$\text{Ⓣ} = \tfrac{1}{2} I_1 I_2 M_{max}[\cos^2\omega_1 t + \cos\omega_1 t \cos(\omega_2 - \omega_3)t]$$

$$= \tfrac{1}{2} I_1 I_2 M_{max}\left(\frac{1 + \cos 2\omega_1 t}{2} + \cos\omega_1 t \cos(\omega_2 - \omega_3)t\right)$$

and nett average torque, $\text{Ⓣ}_{av} = \tfrac{1}{4} I_1 I_2 M_{max}$.

The condition chosen may be stated $\omega_3 = (\omega_1 - \omega_2)$, i.e. the rotational speed is directly related to the frequencies in the two windings. This condition also demonstrates that torque is developed when $\omega_2 = (\omega_1 - \omega_3)$ where ω_2 is the difference speed between the* field of winding 1 and the speed of rotor displacement. It is termed the 'slip speed' and is important in further studies of the induction motor.

* It should be noted that a single phase alternating field can be resolved into two component fields, rotating in opposite directions at synchronous speed, (p. 161). Hence the single phase machine is not self starting and must be spun in the desired direction of rotation. It will run equally well in either direction.

4. The d.c. machine.

When the rotating coil is co-axial with the field axis, ($\theta = 0°$), maximum flux linkage occurs with the coil. On the assumption of a sinusoidal flux distribution in the air-gap, mutual inductance can be expressed in terms of displacement, θ as $M = M_d \cos\theta$ where the suffix refers to the direct-axis of the field, (Fig. 3.6).

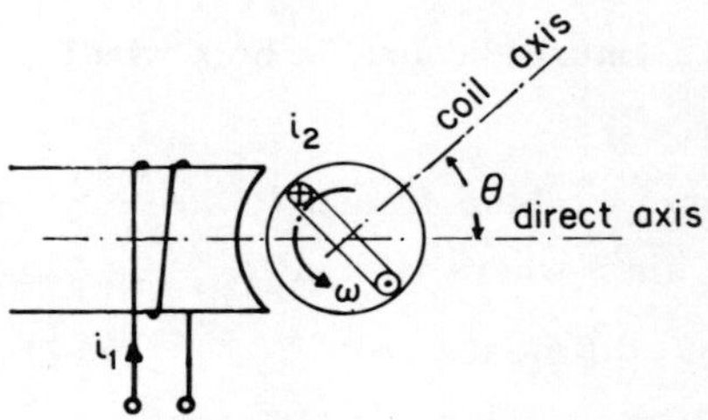

Fig. 3.6.

Assuming, for the moment, L_1 and L_2 are constant:

$$\text{Then} \qquad \text{Ⓣ} = i_1 i_2 \frac{dM}{d\theta}$$

$$= -i_1 i_2 M_d \sin\theta.$$

When $\theta = 90°$ then Ⓣ $= -i_1 i_2 M_d$ (generator action).

When $\theta = -90°$ then Ⓣ $= +i_1 i_2 M_d$ (motor action).

This is the case for the d.c. machine. A continuous, uniformly distributed rotor winding is provided with a commutator and fixed brushes, which maintain $\theta = \pm 90°$. The winding therefore has a consistent configuration with respect to the fixed field axis and to the d.c. polarity fixed brushes (Fig. 3.7). Hence M_d, L_1 and L_2 are constants.

More precise details of the physical features of the machine are considered later.

Note: (a) i_1 may be designated the field current $= i_f$.

i_2 is then called the armature current $= i_a$.

$$\text{Then} \qquad \text{Ⓣ} = M_d i_f i_a$$

$$\omega \text{Ⓣ} = \omega M_d i_f i_a$$

$$= e_a i_a \text{ (neglecting losses).}$$

whence $e_a = \omega M_d i_f$ and is a rotational voltage.

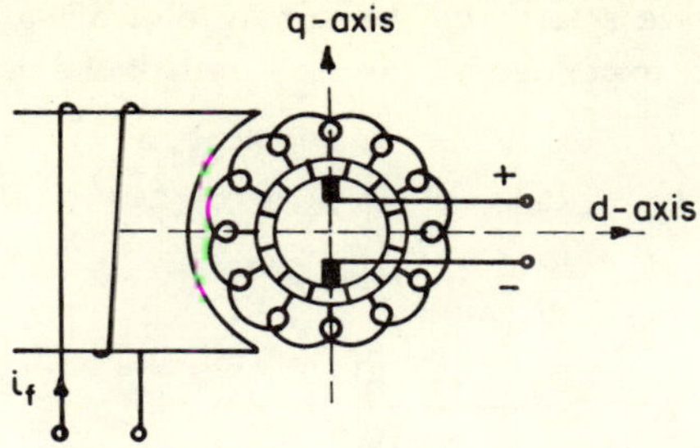

Fig. 3.7.

(b) The rotational voltage is a maximum when the coils are at right angles, ($\theta = 90°$), and zero when the two coils are co-axial, ($\theta = 0°$).

Reference to the voltage equations (p.61), further shows that when the coils are co-axial, only a transformer e.m.f. occurs (i.e. as in a static transformer), while no transformer e.m.f. occurs between coils magnetically at 90°.

(c) The signs on the torque equations above follow the convention that when i_1 and i_2 are positive (supplied to the coils), the torque of electrical origin at the shaft causes rotation in a positive direction.

Torque applied to the shaft (from an external source), is positive if one current is reversed, i.e. a generated current.

REFERENCES

1. WHITE, D.C. and WOODSON, H.H. (1959). *Electromechanical Energy Conversion*. Wiley, New York.
2. MESSERLE, H.K. (1965). *Dynamic Circuit Theory*, Pergamon Press, New York.
3. SKILLING, H.H. (1962). *Electromechanics*, Wiley, New York.
4. FITZGERALD, A.E. and KINGSLEY, C. (1961). *Electric Machinery*, 2nd ed. McGraw-Hill, New York.

EXAMPLES

EXAMPLE 3.1.

An electromagnet in the form of a U has an airgap, between each pole and an armature, of 0·05 cm. The cross-sectional area of the magnet core is 5 cm² and it is uniformly wound with 100 turns. Neglecting leakage and fringing flux, calculate the current

necessary to give a force on the armature of 15 kg. Assume 15% of the m.m.f. is expended on the iron part of the magnetic circuit.
Solution:

$$F = 15\,\text{kg} = 15 \times 9{\cdot}81 \text{ newtons.}$$

$$= \frac{B^2 A}{2\mu_0} \times 2$$

$$= \frac{B^2 (5 \times 10^{-4})}{4\pi \times 10^{-7}}$$

$$\therefore \quad B = 0{\cdot}61\ \text{Wb/m}^2 = \mu_0 H$$

$$H = \frac{0{\cdot}61}{4\pi} \times 10^7\ \text{AT/m}$$

$$\therefore \quad 115\%\ (\text{AT on gaps}) = \frac{1{\cdot}15 \times 0{\cdot}61 \times 10^7 \times 2 \times 0{\cdot}05 \times 10^{-2}}{4\pi}$$

$$= 563{\cdot}5\ \text{AT}$$

$$\therefore \quad I = 563{\cdot}5/100 = 5{\cdot}635\ \text{A.}$$

EXAMPLE 3.2.

Using the principle of field energy density, derive an expression for the torque developed by a singly-excited elementary two-pole reluctance motor, assuming constant-current conditions. Hence calculate the maximum torque developed if the rotor radius is 4 cm, length of air-gap between a pole and the rotor = 0·25 cm, axial length of rotor = 3 cm. The pole excitation is provided by 1000 turns carrying 5 A.
Solution:

There are two airgaps in series, say each = g

$$\therefore \quad B = \mu_0 Ni/2g$$

$$\text{Stored magnetic energy} = (\tfrac{1}{2} B^2/\mu_0)\ \text{volume}$$

$$\text{Volume/gap} = r\theta g l$$

$$\therefore \quad \text{Stored magnetic energy} = 2\left(\frac{1}{2\mu_0} \times \frac{\mu_0^2 N^2 i^2}{4g^2} \times r\theta g l\right)$$

$$= \mu_0 \frac{rl}{4g} N^2 i^2 \cdot \text{d}\theta.$$

Change in electrical input = $ei \cdot \text{d}t = (N\,\text{d}\phi/\text{d}t)\, i\, \text{d}t = Ni \cdot \text{d}\phi$

$$\Phi = rlB\theta = \frac{\mu_0 rl}{2g} Ni\,\theta$$

$$\therefore \text{ Change in electrical input } = \frac{\mu_0 r l}{2g} N^2 i^2 d\theta.$$

Using the energy balance equation,

$$\text{Ⓣ}\, d\theta + \frac{\mu_c r l}{2g} N^2 i^2\, d\theta = \frac{\mu_0 r l}{4g} N^2 i^2\, d\theta$$

$$\therefore \text{Ⓣ} = \frac{\mu_0 r l}{4g} N^2 i^2.$$

Numerical part: in m.k.s. units,

$$\text{Ⓣ} = \frac{4\pi \times 10^{-7} \times 4 \times 10^{-2} \times 3 \times 10^{-2} \times 10^{6} \times 25}{4 \times 0{\cdot}25 \times 10^{-2}}$$

$$= 1{\cdot}2\pi \text{ newton.metres.}$$

4 The electrical machine

Physical features

Details of the physical construction of rotating electrical machines are not only of importance to the designer, but also provide a necessary basis for the understanding of the characteristics of machine performance. In addition, a better perspective is obtained of the several simplifying assumptions commonly adopted in popular methods of analysis.

A rotating electrical machine has two main elements, the stator and the rotor, each of which is manufactured from selected grades of magnetic steel, essentially providing an efficient magnetic circuit. Each carries a winding, which may be termed the field winding, the other being the armature winding. Electrical access to the rotating element or rotor, is by means of fixed contacts termed 'brushes', bearing on rotating 'slip-rings', or perhaps on a 'commutator', Fig. 4.1.

The d.c. machine requires a commutator and hence the rotor carries the armature winding and the commutator. The stator carries inward projecting pole cores which carry the field winding. The a.c. generator and the synchronous motor carry the distributed armature winding on the stator, the poles which form the field system being carried on the rotor. In a.c. machines having uniform airgaps, e.g. the induction motor and the non-salient pole synchronous machine, both stator and rotor carry distributed windings on the airgap periphery.

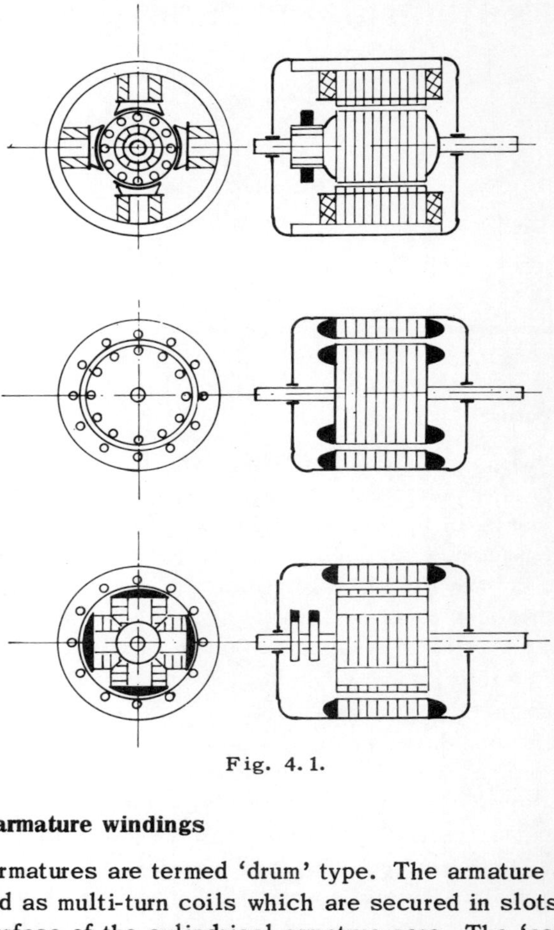

Fig. 4.1.

Typical armature windings

Modern armatures are termed 'drum' type. The armature conductors are wound as multi-turn coils which are secured in slots on the airgap surface of the cylindrical armature core. The 'active' or current-carrying coil-sides may be interconnected in a variety of ways. The title applied to a particular style of winding is determined by the arrangement of the end-connections between the several armature coils. Only the principles of basic types are considered here, since individual variations are a design study.

d.c. machine armature windings

The d.c. or commutator winding is termed a 'double-layer' winding

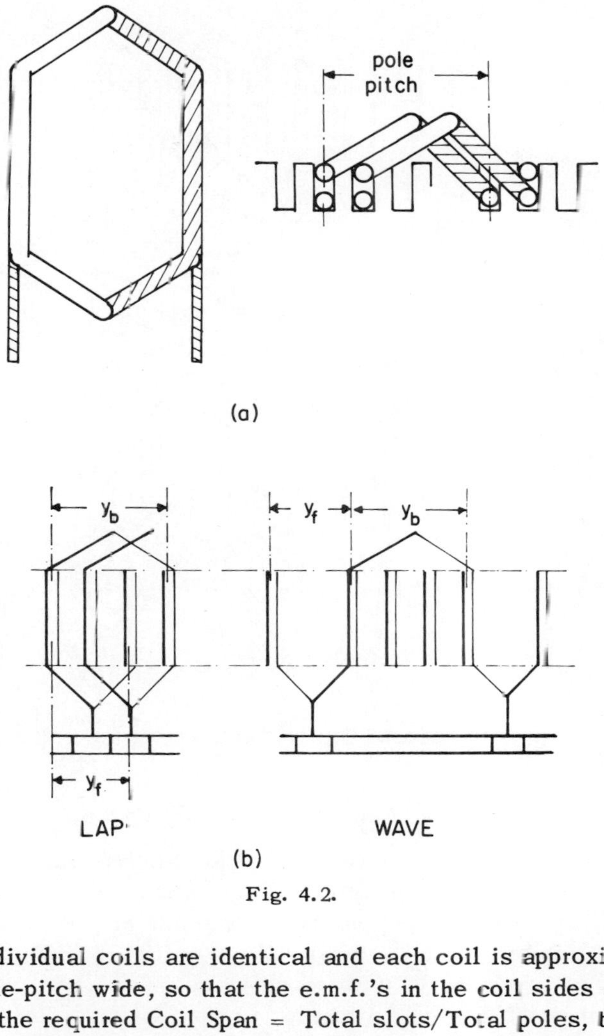

Fig. 4.2.

The individual coils are identical and each coil is approximately one pole-pitch wide, so that the e.m.f.'s in the coil sides add up. Hence the required Coil Span = Total slots/Total poles, to the nearest whole number. One coil-side occupies the bottom of a slot, the other side the top of a slot one pole-pitch away. This results in a characteristic kink in the end-connection, but gives a neat and compact arrangement of end-connections, Fig. 4.2.

The necessary individual coils are first wound on a former and then bound or taped. The prepared coils are located and wedged in the armature slots, resulting in a mechanically robust

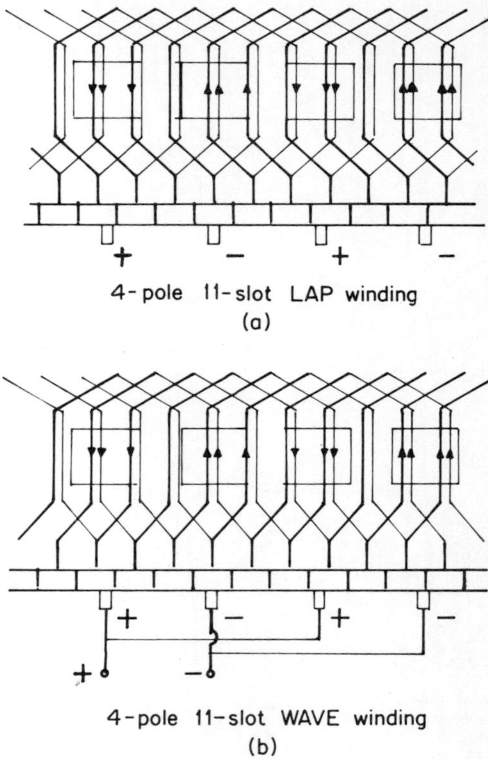

Fig. 4.3.

arrangement. It remains to make the desired inter-coil connections at the 'front' or commutator end of the armature.

The 'front-end' connections are arranged to provide either a LAP or 'parallel' winding, or a WAVE or 'series' winding. To achieve progression, (or retrogression), in the sequence of connections the front-end pitch, y_f, is one slot less, (or more), than the 'back-pitch', y_b, or coil-span. The methods of inter-connection are demonstrated in the developed diagrams, Figs. 4.2(b) and 4.3, and the characteristic 'lap' and 'wave' forms will be recognised.

Following in sequence the directions of the coil-side e.m.f.'s, it will be seen that, for a LAP winding, four points of e.m.f. reversal occur in a four pole winding, Fig. 4.3(a). The brush positions are located on the commutator at these points, corresponding to coil-sides at the inter-polar gaps, i.e. on the quadrature axes.

For a WAVE winding only two points of e.m.f. reversal occur, irrespective of the number of poles and hence only two brushes are necessary. If we consider four brushes located on the WAVE winding, Fig. 4.3(b), as in the LAP case, examination shows that only two coil sides lie between each pair of like-polarity brushes. These coil-sides lie in inter-polar or fringing flux and contribute negligible e.m.f. Hence they act only as a parallelling bridge between similar brushes. One pair of brushes only (+ and –), is necessary but the addition of further brushes as suggested, can reduce the length of the commutator and the current/brush, and this is done on large wave machines. Nevertheless, only two parallel circuits provide the e.m.f. between the brushes and this fact is important in calculating the terminal voltage.

The important conclusion is that:

Number of parallel paths in a LAP winding = Number of poles.

Number of parallel paths in a WAVE winding = two.

Having established the physical form of the winding, the following observations may be made:

(a) The terminal voltage at the brushes will be determined by the number of armature conductors in series between the brushes. The current per brush terminal will be determined by the number of parallel armature circuits. Hence the WAVE winding gives greater voltage, smaller current ratings; the LAP winding gives greater current, smaller voltage ratings.

The construction of the commutator is largely a mechanical matter. If the number of bars is doubled, a second identical armature winding may be arranged in parallel electrically with the first, thereby increasing the number of parallel circuits and the current capacity. This is termed a 'duplex' winding. Multiplex windings may be used for specialist purposes.

(b) Location of the correct brush position on the magnetic neutral axis, (m.n.a.), or q-axis, may be carried out as follows. With the machine at rest a small a.c. is supplied to the field winding. An a.c. voltmeter is connected to the brushes. The brushgear assembly is then moved until a null or minimum reading is obtained on the voltmeter. This result locates the brushes on

the m.n.a., since no transformer e.m.f. occurs between brushes on the q-axis, (Ref. p. 67).
Note that due to geometric symmetry of the end connections, the brushes in the developed diagram are drawn at the centre of a pole. This 'appearance' can occur also in a real machine.
(c) The armature winding forms a closed circuit, but no current flows in it when there is no external circuit connected to the brushes. This may be seen by noting the resultant of the several coil-side e.m.f.s around the total armature circuit. Alternatively, using the 'flux-threading' rule, the total polar flux through the complete armature winding is constant, although the coil revolves, (Fig. 3.7). Hence there is no rate of change of flux, and no e.m.f. around the coil as a whole.

a.c. machine windings

Basically a closed, commutator armature winding is used for an a.c. commutator motor. Indeed it will be seen that a 'series' d.c. motor may run on a.c. since both field and armature currents alternate together, thereby maintaining a unidirectional torque. Commutation problems are more difficult and special refinements can be involved. With three-phase excitation, a synchronously rotating field is produced, (p. 162). Three symmetrically placed brushes per pole pair on the commutator will provide three equal, equally-spaced-in-time a.c. voltages between the brushes, taken in consecutive pairs. The continuous armature winding effectively assumes the form of a delta connection, (p. 83), in the two-pole case. In the four-pole case there would be a second set of three brushes, in parallel with and diametrically opposite to the first set, giving six equally spaced brushes in all.

a.c. machine 'open' windings.

a.c. machine windings resemble d.c. windings. But they are fundamentally different when they are 'open' windings, i.e. both ends are brought out for each phase winding, instead of being closed as in d.c. and a.c. commutator windings.

Each coil should be one pole-pitch in span for maximum e.m.f., but this can be varied to improve waveform. They may be single-layer or double-layer windings. The single-layer arrangement has one coil-side per slot. The double-layer arrangement has two coil-sides in each slot. Typical examples are shown in

Figs. 4.4 and 4.5.

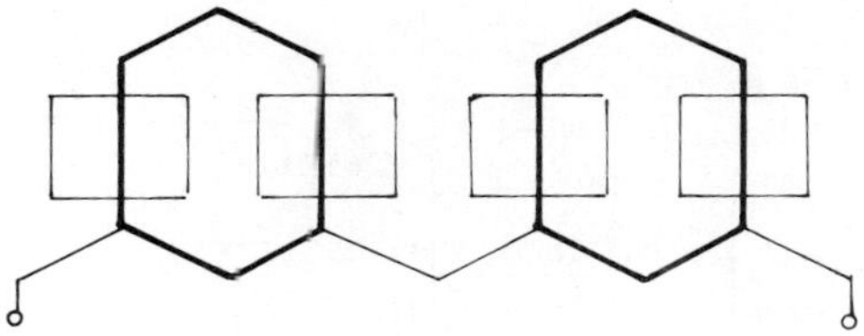

(a) concentrated

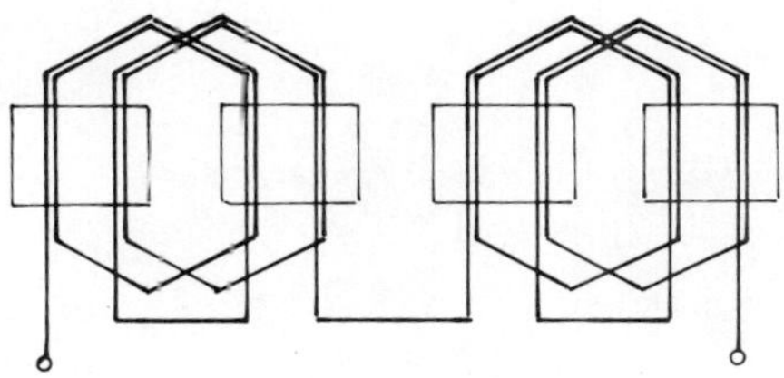

(b) distributed hemitropic

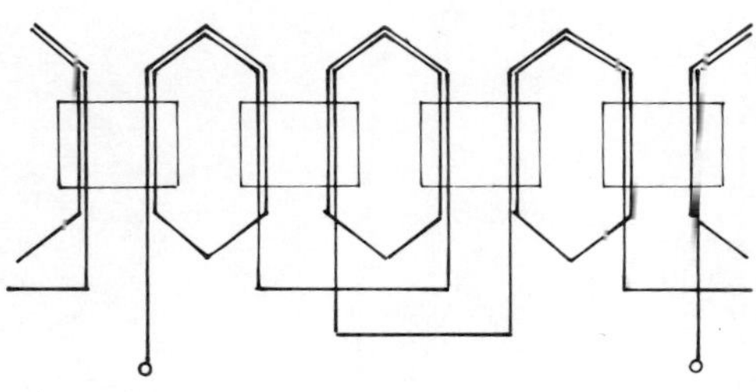

(c) distributed

Fig. 4.4.

Principles of a.c. windings

Consider the simple a.c. generator represented in Fig. 4.6. A single coil aa' would provide a single-phase alternating voltage. Since one cycle is generated per pair of poles on the rotor then, in general:

$$\text{frequency, } f = \frac{pN}{60} \text{ c.p.s. where } N = \text{r.p.m.}$$

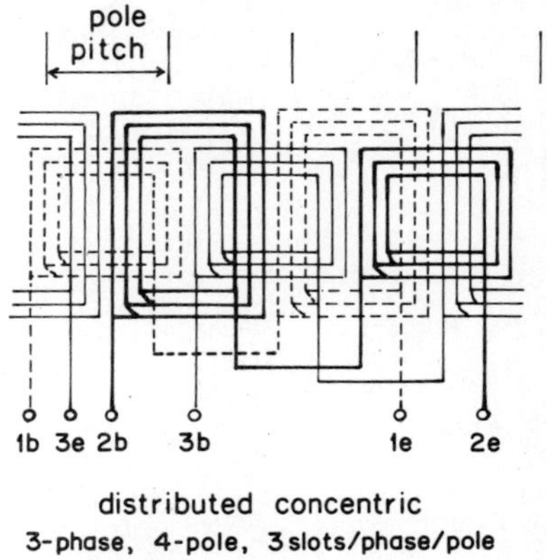

Fig. 4.5.

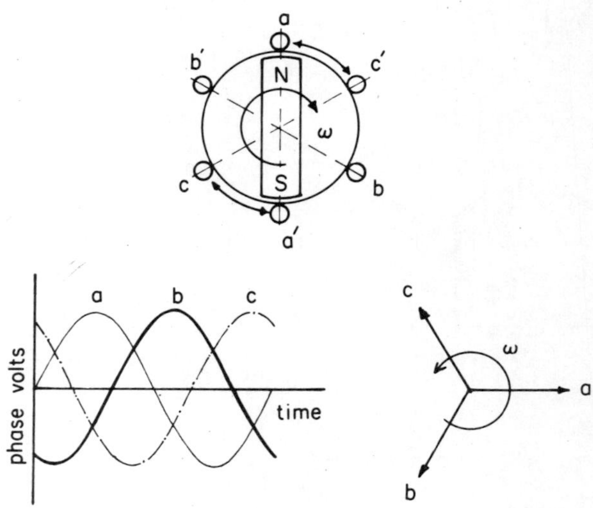

Fig. 4.6.

Further coils, bb' and cc', would provide identical single-phase voltages except that, on a time scale, they would all be spaced 1/3 cycle apart due to the 120° spacing of the coils, Fig. 4.6. The result is a simple three-phase generator.

Since many turns are required in a phase winding to generate a useful voltage, either very deep slots would be required or the winding may be distributed to occupy several slots within the phase spread, a–c', etc. (Fig. 4.6). Deep slots are costly, whereas advantages attend the distribution of the phase winding, namely (*i*) better utilisation of the available stator surface and better cooling, (*ii*) improvement of waveform.

Distribution factor, k_d.

The phase voltage will be the phasor sum of the coil voltages. Consider the general case of 'm' full-pitch coils/phase distributed in slots spaced at $a°$ electrical. The coil voltages form the sides of the polygon-shaped phasor diagram for the resultant phase voltage. (Fig. 4.7 (a)).

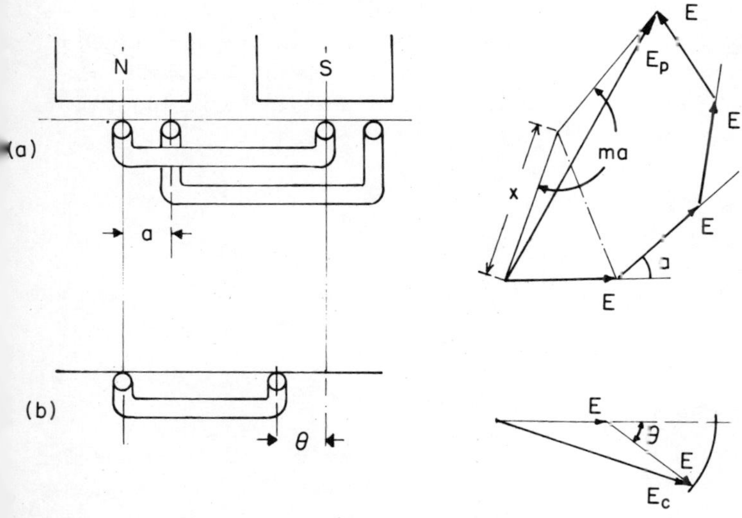

Fig. 4.7.

Let E = volts/coil.

Resultant volts/phase, $E_p = 2x \cdot \sin \frac{ma}{2}$.

But $x = \dfrac{E/2}{\sin \frac{a}{2}}$

$$E_p = \frac{E \sin \frac{ma}{2}}{\sin \frac{a}{2}} .$$

The ratio $\frac{\text{Phasor sum}}{\text{Arithmetic sum}}$ is termed Winding Distribution Factor, k

Hence:

$$k_d = \frac{E_p}{mE} = \frac{\sin \frac{ma}{2}}{m \sin \frac{a}{2}}$$

where a = slot pitch in electrical degrees,

m = coils/phase,

= slots/phase/pole, (sometimes remembered as spppp).

Reference to the phasor diagram and this equation shows that excessive spread of the phase winding adds little to the magnitude of the generated voltage.

Space harmonics in the flux density wave are interpreted in terms of 'harmonic poles' with respect to the slot angle, in electrical degrees. Thus the distribution factor for the harmonic voltage component will be:

$$k_{dh} = \frac{\sin \frac{nma}{2}}{m \sin \frac{na}{2}}$$

for the nth harmonic. If 'a' can be made very small, for example by skewing the slots, higher-order harmonics can be virtually eliminated.

Coil span factor, k_c.

If the coils are not full-pitch then they are either short-chorded or long-chorded. As above, the voltage/coil will be the phasor sum of the voltages of the coil-sides.

The ratio $\frac{\text{phasor sum}}{\text{arithmetic sum}}$ = Coil Span Factor, k_c.

Reference to the simple phasor diagram, (Fig. 4.7 (b)) shows:

$$\frac{E_c}{2E} = \cos \frac{\theta}{2} = k_c$$

where E = voltage/coil-side, E_c = voltage/coil and $\theta°$ = angle of departure from full-pitch of the coil-span.

Harmonics are again interpreted in terms of 'harmonic poles' and the appropriate angle in electrical degrees is $n\theta/2$. Thus for the nth harmonic:

$$\text{Coil Span Factor, } k_{cn} = \cos\frac{n\theta}{2}.$$

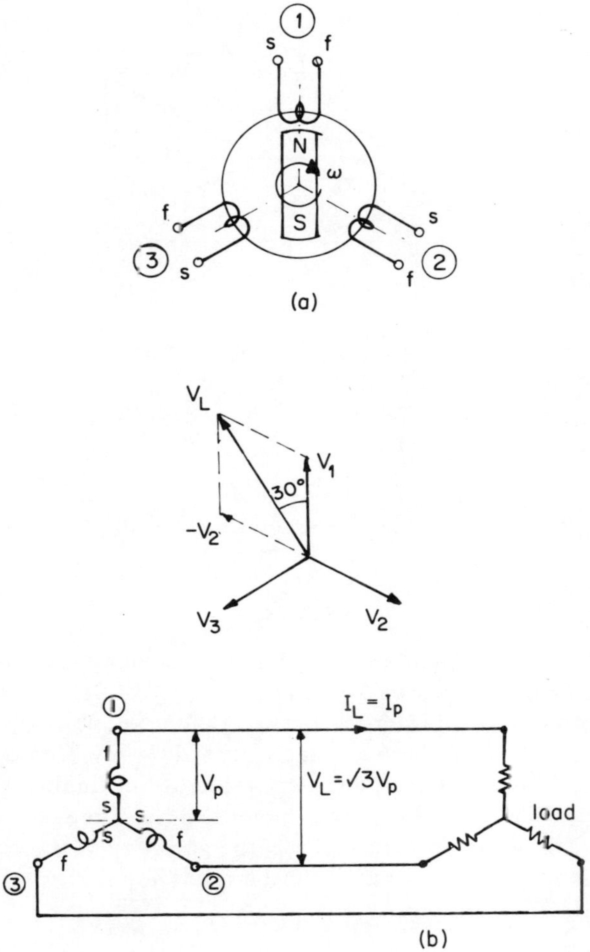

Fig. 4.8.

Hence by choosing θ so that k_{cn} becomes zero, a given harmonic can be eliminated. For example, 3rd harmonics are eliminated when $\theta = 60°$, that is coil span $= 120°$, or 2/3 full-pitch.

Interconnection of phase windings

Advantage is taken of the availability of the three phase supplies, in respect of voltage or of current and of the greater power. A three-phase machine is more economical and of lesser bulk than the equivalent three single-phase machines.

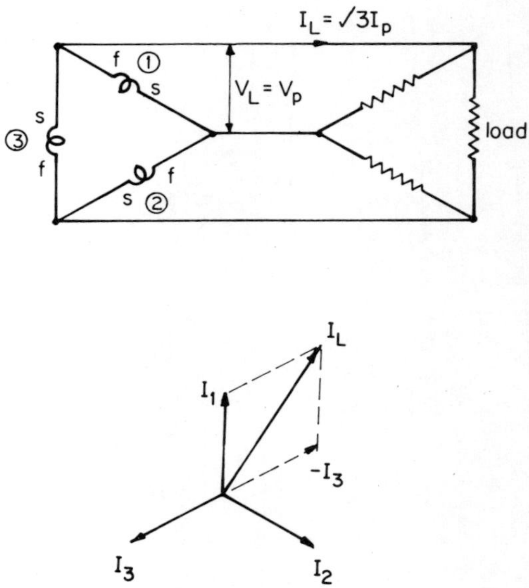

Fig. 4.9.

The three phase windings and the corresponding three phase-voltage phasors are shown schematically (Fig. 4.8). In the Star or Wye, (Y) method of interconnection, the three 'start' (s) points are commoned. Then three voltages, termed the Line voltages, are available one at each pair of 'finish' (f), terminals. The common point is termed the 'star' point and the three output terminals, the Line terminals. From the geometry of the phasor diagram it will be seen that, (note the relative polarities),

$$\text{Line voltage, } V_L = \sqrt{3}\, V_p$$

and

$$\text{Line current, } I_L = I_p \, .$$

In the Delta, (Δ), method of interconnection, the phases are connected in series, (s) to (f) in sequence, (Fig. 4.9). Noting the relative polarities and with a balanced three-phase loading the phasor diagram shows:

$$\text{Line current, } I_L = \sqrt{3}\, I_p$$
$$\text{Line voltage, } V_L = V_p\,.$$

In both cases, three-phase power = 3 (power/phase)

$$= 3\, V_p\, I_p \cos\phi$$
$$= \sqrt{3}\, V_L I_L \cos\phi.$$

Also, note that the resultant voltage around the Delta is zero, while the 'neutral' or star-point current is zero with a balanced three-phase loading.

General statement of the e.m.f. equation

Electromagnetic induction may be defined simply:

$$e = \frac{d\lambda}{dt} = N \cdot \frac{d\phi}{dt}$$

where N = effective turns linked.

Assuming sinusoidal distribution of flux in the airgap:

$$\phi = \Phi \cos\theta.$$

Both Φ and θ can be functions of time, θ due to rotation and $\theta = \omega t$, Φ due to an a.c. field supply.

$$\text{Hence, } e = N\frac{d\Phi}{dt}\cos\omega t - \omega N\Phi \sin\omega t$$
$$= \text{Transformer volts} + \text{Rotational volts.}$$

If the flux/pole is pulsating, in general both transformer and rotational e.m.f.s will occur. If the flux/pole is constant, i.e. d.c. excited, there will be only a rotational voltage. Thus, for an a.c. synchronous generator:

$$e_{rot} = \omega N\Phi \sin\omega t$$
$$E_{max} = \omega N\Phi = 2\pi f N\Phi$$
$$E_{rms} = \frac{1}{\sqrt{2}} E_{max} = \frac{2\pi}{\sqrt{2}}\,\Phi f N$$
$$= 4{\cdot}44\,\Phi f N.$$

This equation has the same form as that obtained for the e.m.f. of a static transformer. But it is different in that it is a rotational voltage, whereas the transformer equation is derived from the differential of a sinusoidally pulsating flux and is a transformer e.m.f.

For an a.c. machine it is convenient to express the e.m.f. in terms of armature conductors. Thus turns, $N = Z_p/2$ where Z_p = series conductors/phase winding.

$$\therefore \quad E_{rms}/\text{phase} = 2{\cdot}22\,\Phi f Z_p .$$

Allowing for distribution of the winding we include a 'winding factor', $k_w = k_d k_c$, giving:

$$E_{rms}/\text{phase} = 2{\cdot}22\,k_w\,\Phi f Z_p \text{ volts.}$$

Field form, e.m.f. and torque of a d.c. machine

The distribution of flux density in the air-gap of a d.c. machine, displayed graphically, is termed the field form (Fig. 4.10). This two-dimensional representation is valid on the assumptions that axial distribution of flux density is uniform and 'end effects' are negligible.

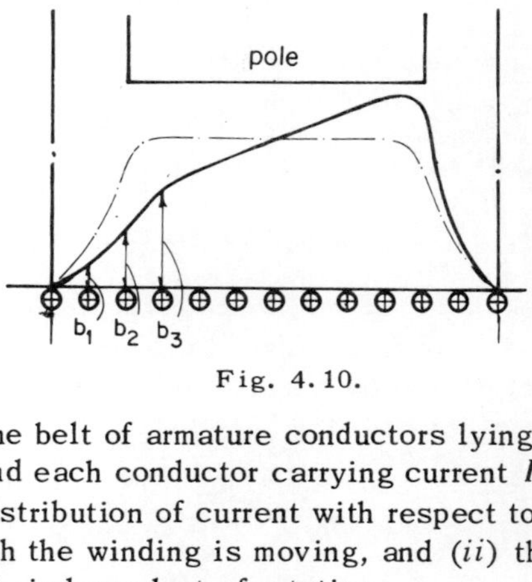

Fig. 4.10.

Consider the belt of armature conductors lying within the field of one pole and each conductor carrying current I_c. Remember (*i*) that the distribution of current with respect to the brushes is static although the winding is moving, and (*ii*) that the torque is a static effect, independant of rotation.

Torque due to each conductor located in a local field density, b:

$$\text{Ⓣ}_c = b l I_c r .$$

Torque developed by the several conductors in each pole pitch, $\text{Ⓣ}_p = l I_c r\,(b_1 + b_2 + b_3 + \ldots)$.

If x = conductor spacing and Z_p = conductors/pole-pitch.

Then $x \cdot Z_p = 2\pi r/2p$

and $$Ⓣ_p = \frac{l I_c p Z_p}{\pi}[x(b_1 + b_2 + b_3 + \dots)].$$

By the trapezium rule, the area under the field form is given by the expression in the brackets, [] = A, and

$$\text{Flux/pole} = Al = \Phi.$$

Thus $$Ⓣ_p = \frac{p\Phi Z_p I_c}{\pi}.$$

For an armature with 'a' parallel paths, terminal current $I = aI_c$. Hence, in terms of total conductors Z and terminal current I:

$$Ⓣ_p = \frac{p\Phi Z I}{2\pi p a} = \frac{\Phi Z I}{2\pi a}.$$

Hence total torque, for $2p$ poles:

$$Ⓣ = \frac{p\Phi Z I}{\pi a} \text{ newton.metres.}$$

For the d.c. machine we note that the voltage in a given armature conductor will be alternating, but the action of the commutator is to provide full-wave rectification. Thus the d.c. value of the voltage between the brushes is the average. Assuming sinusoidal flux-distribution and a concentrated coil of $Z_s/2$ turns:

$$e_{d.c.} = \frac{1}{\pi}\int_0^{\pi} \omega \cdot \frac{Z_s}{2}\Phi \cdot \sin\theta \cdot d\theta$$

$$= \frac{2}{\pi}\omega \cdot \frac{Z_s}{2}\Phi$$

$$= 2\Phi f Z_s \text{ where } Z_s = \text{series armature conductors.}$$

Alternately, from the previous section:

$$E_{av} = E_{rms}/1{\cdot}11 = 2\Phi f Z_s$$
$$= \frac{2p\Phi N Z}{60a} = E_{d.c.}$$

where a = parallel paths in the armature winding, N = r.p.m. This result is correct but the matter is further examined to justify the validity of the idealised assumptions above.

A d.c. motor running at N r.p.m. has an armature current I, and develops a gross torque Ⓣ. If E = the part of the supply voltage equivalent to the armature 'back' voltage, then equating the power developed by the armature to the supply power:

$$EI = \omega \text{Ⓣ} = \frac{2\pi N \text{Ⓣ}}{60}$$

$$= \frac{2\pi N}{60} \frac{p\Phi Z I}{a\pi}$$

$$\text{Hence} \quad E = \frac{2p\Phi N Z}{60a} \text{ volts.}$$

An identical expression is obtained for the generated voltage of a d.c. generator, driven at N r.p.m. and delivering I amps, again by equating $EI = \omega$ Ⓣ, and substituting as before.

It will be seen later, under the title of Armature Reaction, that the field form is distorted by the field caused by current flowing in the armature conductors. If the resultant field form were used, the same results would be obtained, if the total flux/pole is unchanged. But calculation of the force on an individual conductor would be changed, since it would depend on the resultant local B, due to the pole and the neighbouring current-carrying conductors.

Alternative proof of d.c. voltage equation

The voltage generated in a single conductor of length 'l' moving in a local field density 'b':

$$e = blv = bl\omega r.$$

If the e.m.f.s of the total armature conductors, Z, were additive,

$$\text{Total volts generated} = e_t = l\omega r(b_1 + b_2 + b_3 + \ldots).$$

If x = conductor spacing, then $Zx = 2\pi r$ and $r = Zx/2\pi$:

$$e_t = \frac{l\omega Zx\,(b_1 + b_2 + b_3 + \ldots)}{2\pi}.$$

Since all the conductors were assumed in series:

$$lx\,(b_1 + b_2 + b_3 + \ldots) = \text{total gap flux} = 2p\Phi.$$

$$\text{Then} \quad e_t = \frac{\omega Z \cdot 2p\Phi}{2\pi} = \frac{2\pi N}{60} \cdot \frac{2p\Phi Z}{2\pi} = \frac{2p\Phi ZN}{60}.$$

But there are 'a' parallel groupings of the armature conductors and the conductors in series $= Z/a$.

Hence, as before, $$E = \frac{2p\Phi NZ}{60a} \text{ volts.}$$

These expressions show that the torque and e.m.f. of a d.c. machine, with brushes located on the quadrature axis, are:
(a) independent of the space distribution of flux and depend on the total flux per pole,
(b) independent of the distribution of the series armature conductors and the armature winding may thus be interpreted as an equivalent concentrated coil.

Factors in design

The designer employs available materials to best advantage for the manufacture of an efficient machine. In this he will be guided to a great extent by certain aspects of previous successful design practice.

Limitations are imposed by magnetic saturation, by hysteresis and eddy current losses in the iron parts and by ohmic resistance losses in the windings. The physical dimensions and the speed are both important factors in relation to dissipation of the heat losses.

The magnetic and the electric loadings of the machine are commonly defined as follows:
(a) *the specific magnetic loading*, or mean working flux density in the air gap, is expressed:

$$\overline{B} = \frac{2p\Phi}{\pi DL}.$$

(b) *the specific electric loading*, or ampere-conductors per metre of armature periphery, is expressed:

$$\overline{ac} = \frac{ZI_c}{\pi D}.$$

The product, D^2Ln is proportional to the volume enclosed by the air gap surface and to the speed.
Hence, for the d.c. machine:

$$E = \frac{2p\Phi nZ}{a}.$$

The volt-amp rating, $EI = \dfrac{2p\Phi nZI}{a}$

$$= \frac{2p\Phi}{\pi DL} \cdot \frac{ZI_c}{\pi D} \cdot \pi^2 D^2 Ln$$

$$= \pi^2 D^2 Ln \cdot \overline{B} \cdot \overline{ac}.$$

The ratio: $\dfrac{EI}{D^2 Ln} = \pi^2 \overline{B}\, \overline{ac} = G$

is termed the <u>Output Coefficient</u>, and has dimensions joules/m³.

The specification of the machine will decide the speed, n, and with guidance on suitable values for $\overline{B}$ and $\overline{ac}$, the D^2L product is determined. A square pole may be chosen and since pole arc/pole pitch ≏ 2/3:

$$L = \frac{2}{3} Y = \frac{2}{3} \frac{\pi D}{2p}$$

$$D^2 L = \frac{\pi D^3}{3p}.$$

However, modern designs tend towards small diameters and long cores. Modern developments have led to improved magnetic materials and heat resisting insulating materials permitting smaller machines for a given power.

The choice of speed in a.c. machines is related to frequency, and in d.c. machines to the type of prime-mover. In large machines, peripheral speed can be a limiting factor due to centrifugal forces. Hydro-electric machines are designed with respect to the 'run-away' speed.

The Output Coefficient for the three-phase a.c. machine is determined as follows:

$$E_p = 2{\cdot}22\, k_w \Phi f Z_p$$

$$\triangleq 2{\cdot}12\, \Phi f Z_p$$

kVA rating, $Q = 3 \cdot E_p I_p \cdot 10^{-3}$

Now $\overline{ac} = \dfrac{3 Z_p I_p}{\pi D}$

$$\overline{B} = \frac{2p\Phi}{\pi DL}.$$

Then $Q = \dfrac{2{\cdot}12}{2} \pi^2 \cdot D^2 Ln \cdot \overline{B} \cdot \overline{ac} \cdot 10^{-3}$

$$= 10{\cdot}4\, D^2 Ln \cdot \overline{B} \cdot \overline{ac} \cdot 10^{-3}$$

$$G = \frac{Q \cdot 10^3}{D^2 Ln} = 10 \cdot 4\, \bar{B} \cdot \overline{ac}. \text{ joules/metre}^3.$$

Typical values for $\bar{B}$ range, from F.H.P. machines to large turbo machines, from $0 \cdot 3$ to $0 \cdot 55\ \text{Wb/m}^2$ and for $\overline{ac}$ from 10^4 to $5 \cdot 5 \times 10^4$ ac/m. The working value for the flux density is greatest where the magnetic path is most constricted, namely in the armature tooth, and may rise as high as $2 \cdot 2\ \text{Wb/m}^2$.

Efficiency and losses

The efficiency of any machine is defined:

$$\eta = \frac{\text{Output power}}{\text{Input power}}.$$

Thus for a d.c. generator, $\eta_g = \dfrac{\text{output}}{\text{output} + \text{losses}}$

$$= \frac{VI}{VI + \text{losses}}.$$

For a d.c. motor, $\eta_m = \dfrac{\text{bhp} \times 746}{\text{input watts}}$ or, $\eta_m = \dfrac{VI - \text{losses}}{VI}$.

The losses incurred in a machine are of the same nature for both generator or motor operation. It will be seen that they are independent of the mode of operation of the machine. The losses are defined as follows:

(*a*) *Rotational losses.*

(*i*) *mechanical losses.* These are known as the friction and windage losses and are a function of speed. Friction losses occur mainly due to brush friction on the commutator.

(*ii*) *core losses, due to hysteresis and eddy-current effects.* The fundamental nature of these losses is discussed in Chapter 3, where it is shown:

$$W_{\text{core}} = k_h B^x f + k_e B^2 f^2 \ \text{W/m}^3$$

and x = the Steinmetz index.

(*b*) *Electrical losses.*

(*i*) *copper loss in the armature winding* $= I_a^2 R_a$ watts, and varies with load current. Accurate measurement of R_a can involve using probes under the brush contacts to measure the volt drop on R_a for a given current. Allowance may be made for brush drop on an arbitary basis of two volts total.

(*ii*) *copper loss in the field winding* = $I_f^2 R_f$ watts, and for a shunt-connected field, is a constant.

Condition for maximum efficiency:

Consider the generator, where $\eta_g = \dfrac{VI}{VI + I^2R + C}$

where C = constant losses. In a 'constant voltage' world, load current and load are proportional. To obtain the condition for maximum η,

$$d\eta/dI = \frac{(VI + I^2R + C)V - VI(V + 2IR)}{(\text{denominator})^2} = 0.$$

Then $I^2R = C$, or Variable loss = Constant loss, is the condition, and is a general law for all machines.

The Hopkinson efficiency test

This test is an important example of 'regenerative', or 'back-to-back' tests. Two similar, preferably identical, d.c. shunt machines are employed. The machines are mechanically coupled, one to act as a motor driving the other as a generator. The electrical output of the generator is fed back to the motor supply, care being taken that the voltages are correct.

Variation of the field excitations of the machines enables full-load conditions to be obtained. The power supplied from the mains is only that necessary to supply the losses of both machines. The method is economical where tests of long duration, or 'heat runs' are required. Calculation of efficiency may be carried out as follows:

(*a*) *Approximate solution for efficiencies.*

If it is assumed that both copper and core losses may be divided equally between the two machines (see Fig. 4.11):

$$\text{Motor input} = VI_3 = V(I_1 + I_2)$$
$$\text{Motor output} = \eta_m V(I_1 + I_2)$$
$$\text{Generator output} = VI_2$$
$$\text{Generator input} = VI_2/\eta_g$$

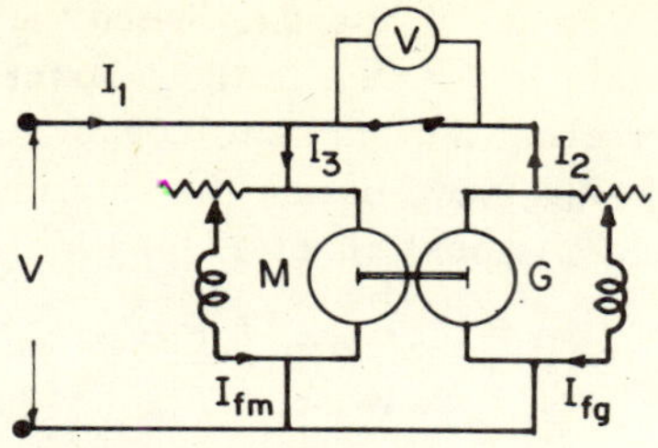

Fig. 4.11.

Hence we have, $$\eta_m \eta_g = \frac{I_2}{I_1 + I_2}$$

and assuming equal efficiencies, $$\eta = \sqrt{\left(\frac{I_2}{I_1 + I_2}\right)}.$$

The answer obtained is approximate, particularly for small machines mainly due to the error in assuming equal armature copper losses.

(*b*) *Solution for individual efficiencies.*

The individual losses may be calculated separately and a small but acceptable error remains in assuming equal rotational losses. The calculation becomes:

Total set input $= VI_1$

Motor arm. copper loss $= (I_3 - I_{fm})^2 R_{am}$

Motor field copper loss $= I_{fm}^2 R_{fm}$

Generator arm. copper loss $= (I_2 + I_{fg})^2 R_{ag}$

Generator field copper loss $= I_{fg}^2 R_{fg}$

$\therefore$ Rotational losses/machine $= \frac{1}{2}$ (set input − total copper losses)

Then, Motor efficiency $= \dfrac{\text{motor input} - \text{motor losses}}{\text{motor input}}$

$= \eta_m$

Generator efficiency, $\eta_g = \dfrac{\text{generator output}}{\text{generator output} + \text{generator losses}}$

Separation of losses by test

In this test the machine is run light as a motor, at various fixed field currents over a range of speeds. Speed variation is obtained by means of a variable resistance in the armature circuit, Fig. 4.12. By subtracting the armature copper loss, $(I_a^2 R_a)$, from the armature input, $(V_a I_a)$, we obtain the rotational losses, W_r. Thus, borrowing from the next chapter, (p. 107), we have:

$$V_a = E_b + I_a R_a$$

$$\text{or,} \quad E_b = V_a - I_a R_a.$$

Then $E_b I_a = V_a I_a - I_a^2 R_a = W_r$, the rotational loss.

R_a is determined by measurement on the machine while it is still warm.

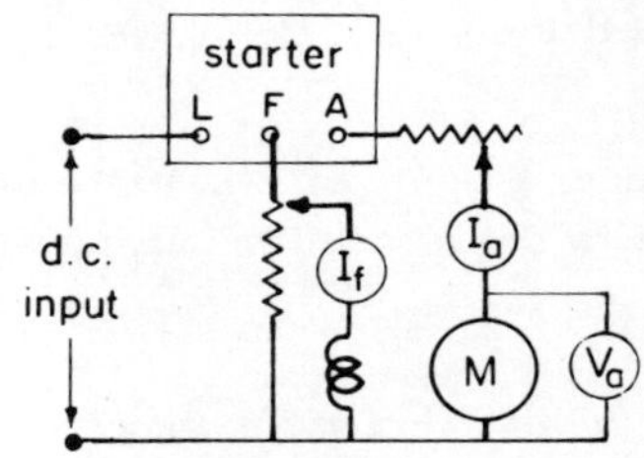

Fig. 4.12.

A family of curves is obtained by plotting W_r against speed, for each constant value of field current, Fig. 4.13. A second family of curves, of W_r/excitation at constant speeds, is derived by taking ordinates at selected speed values.

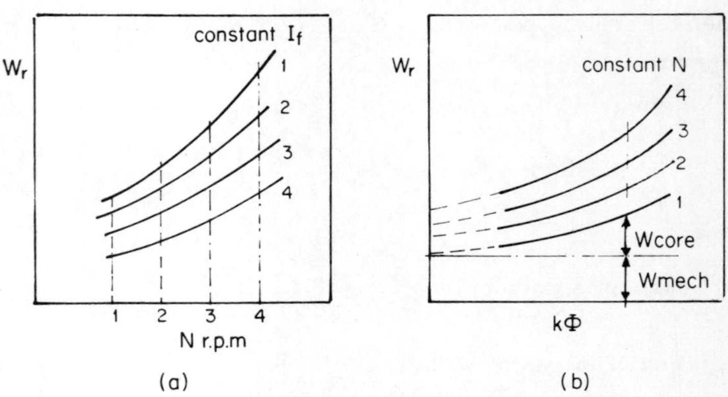

Fig. 4.13.

Now, it is assumed that extrapolation of each curve back to zero excitation gives the mechanical loss corresponding to each constant speed, and this loss is assumed constant over the range of excitation, the difference with W_r being the core loss due to excitation.

N.B. A more accurate estimate of 'zero excitation' may be obtained by replacing each 'constant' I_f value by the corresponding average $k\Phi$ ($= E_b/n$) value, when plotting the second family of curves.

Hence graphs of mechanical loss/speed and core loss/speed (at a selected excitation ordinate), may be derived. Subsequent efficiency of the machine may be predicted, for motor or generator operation, for a range of arbitary loads and using rotational losses measured in this test.

The retardation test

This is a convenient and important test which enables the moment of inertia and the rotational losses of any rotating mass to be determined. The following description applies basically to the case of electrical machines. It will be seen, however, that it can be applied to rotating masses in general.

The motor is run up to a speed above normal and then disconnected from the supply, so that it will run down. The running down is repeated with an electrical load connected to the machine terminals. Observations are made of speed at noted time intervals, and of the electrical load at a given speed (usually normal rated).

$$\text{In general, energy stored in rotor} = \tfrac{1}{2} J\omega^2$$

$$\text{Rate of change of energy} = J\omega \frac{d\omega}{dt}$$

$$= J \frac{2\pi N}{60} \frac{d}{dt}\left(\frac{2\pi N}{60}\right)$$

$$\therefore \quad P = 0{\cdot}011\, JN \frac{dN}{dt}.$$

In test (1) retardation occurs for the machine, excited, but with no connected load.

Then, P_1 = friction and core losses.

$$\text{Slope of speed/time curve at } N = \frac{dN_1}{dt}.$$

In test (2) retardation occurs for the machine, excited but with a known connected load, W.

Then, P_2 = friction and core losses + W.

$$\text{Slope of speed/time curve at } N = \frac{dN_2}{dt}.$$

$$\text{Hence metered load, } W = P_2 - P_1 = 0{\cdot}011\, JN\left(\frac{dN_2}{dt} - \frac{dN_1}{dt}\right).$$

In the following example, rated N = 1000 r.p.m.

$$\text{Test results gave, } dN_2/dt - dN_1/dt = 60{\cdot}8 - 31{\cdot}0 = 29{\cdot}8 \text{ r.p.m./sec.}$$

Metered load at 1000 r.p.m., W = 788 watts.

$$\text{Hence } J = \frac{788}{29{\cdot}8 \times 0{\cdot}011 \times 1000} = 2{\cdot}4\,\text{kg.m}^2 .$$

Also, friction and core losses, $P_1 = 11 \times 2{\cdot}4 \times 31 = 820$ watts.

Where running down speed is rapid, a chart recorder or calibrated tachometer may be employed. Inertia and losses for a non-electric rotor are obtained by coupling to a machine as above, and repeating the test.

REFERENCES

1. MOULLIN, E.B. (1955). *Electromagnetic Principles of the Dynamo* Oxford University Press, London.
2. CLAYTON, A.E. and HANCOCK, N.N. (1961). *The Performance and Design of Direct-current Machines*, Pitmans, New York.
3. FITZGERALD, A.E. and KINGSLEY, C. (1961). *Electric Machinery*, 2nd ed., McGraw-Hill, New York.
4. SAY, M.G. (1961). *Performance and Design of a.c. Machines*, Pitmans.
5. HINDMARSH, J. (1965). *Electrical Machines*, Pergamon Press, Oxford.
6. HUGHES, E. (1965). *Electrical Technology*, Longmans, New York.

EXAMPLES

EXAMPLE 4.1.

The air-gap flux distribution of a three-phase, 50 c/s 1000 r.p.m. alternator contains 3rd and 5th harmonics of magnitudes 10% and 5% of the fundamental respectively.

Calculate the instantaneous voltage/phase generated, if the machine has a single-layer stator winding housed in 72 slots with four conductors/slot. The coil span is 120° elect. and the fundamental flux/pole is 0·05 Wb.

Solution:

$$E_{\text{phase}} = 2{\cdot}22\, k_d k_c \Phi f Z$$

$$\text{Conductors/phase} = Z = 72 \times 4/3 = 96.$$

For the fundamental:

$$k_d = \frac{\sin \frac{ma}{2}}{m \sin a/2}$$

$$m = \text{slots}/ph/\text{pole} = \frac{72}{3 \times 6} = 4$$

$$a = \text{slot pitch} = \frac{180}{12} = 15^\circ$$

$$\therefore\ k_d = \frac{\sin 30^\circ}{4 \times \sin 7{\cdot}5^\circ} = 0{\cdot}958$$

$$k_c = \cos\theta/2 = \cos\frac{(180^\circ - 120^\circ)}{2} = 0{\cdot}866$$

$$\therefore\ \text{Fundamental } E_{ph} = 2{\cdot}22 \times 0{\cdot}958 \times 0{\cdot}866 \times 0{\cdot}05 \times 50 \times 96$$

$$= 441\,\text{V}.$$

$$\text{Max. } E_{ph} = \sqrt{2} \times 441 = 624\,\text{V}.$$

For the third harmonic:

$$\text{With coil span} = 120^\circ,\ \theta = (180^\circ - 120^\circ) = 60^\circ$$

$\therefore\ k_{c_3} = \cos n\theta/2 = 0$, and the 3rd harmonic is eliminated.

For the 5th harmonic:

$$k_{d_5} = \frac{\sin mna/2}{m \sin na/2} = \frac{\sin 150^\circ}{4 \times \sin 37{\cdot}5^\circ} = 0{\cdot}2055$$

$$k_{c_5} = \cos n\theta/2 = \cos 150^\circ = -0{\cdot}866$$

$$\therefore\ E_{5/ph} = 2{\cdot}22 \times 0{\cdot}2055 \times (-0{\cdot}866) \times \frac{0{\cdot}05}{100} \times 5 \times 250 \times 96$$

$$= -23{\cdot}7\,\text{V}$$

$$\text{Max. } E_{5/ph} = -33{\cdot}5\,\text{V}$$

Hence inst. voltage/phase, $\underline{e = 624 \sin\omega t - 33{\cdot}5 \sin 5\omega t.}$

EXAMPLE 4.2.

The power required to drive, at rated speed on open-circuit, a separately-excited machine when unexcited is 600 W, and when excited is 1800 W. At half rated speed the corresponding powers are 300 W and 800 W. Determine the components of the core loss at rated speed.

Solution:

At rated speed: mechanical losses = 600W

core losses = 1800 − 600 = 1200W.

At half rated speed:

mechanical losses = 300W

core losses = 800 − 300 = 500W.

Now

$$W_h \propto f \propto n$$

$$W_e \propto f^2 \propto n^2$$

$$\therefore\ W_h + W_e = 1200$$

$$\tfrac{1}{2} W_h + \tfrac{1}{4} W_e = 500.$$

Solving gives:

$$W_e = 400\text{W}$$

$$W_h = 800\text{W}.$$

EXAMPLE 4.3.

Prove, with suitable diagrams, that two wattmeters may be used to measure the power supplied to a balanced three-phase load if the star-point is not available. A 20 h.p. 400 V three-phase a.c. motor on full load has efficiency of 0·88 p.u. and a power factor of 0·84 lagging. Determine the wattmeter readings if the two-wattmeter method is used to meter the input power to the motor on full load.

Solution:

This is an important method commonly used in metering three-phase power. The wattmeters are arranged as shown in

Fig. Ex. 4.3 (a) and the appropriate phasor diagram in Fig. Ex. 4.3 (b). For convenience of discussion, a star-connected load is shown.

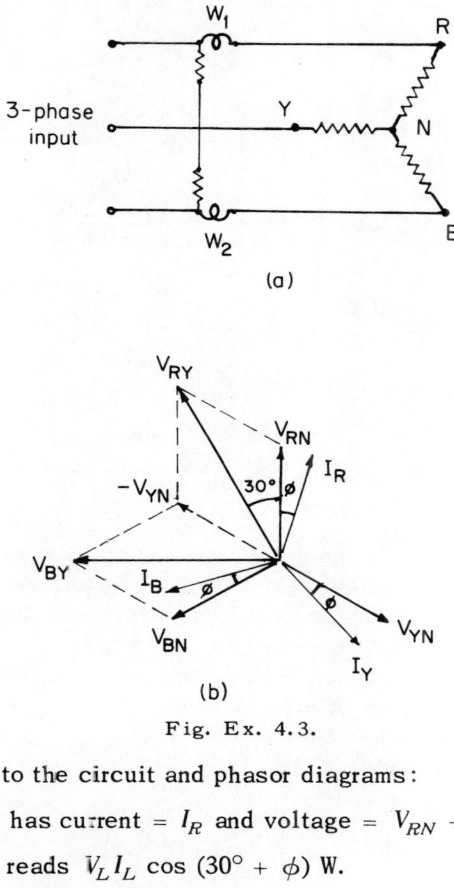

Fig. Ex. 4.3.

Referring to the circuit and phasor diagrams:

Wattmeter W_1 has current $= I_R$ and voltage $= V_{RN} + V_{NY} = V_{RY}$

Hence W_1 reads $V_L I_L \cos(30° + \phi)$ W.

Wattmeter W_2 has current $= I_B$ and voltage $= V_{BN} + V_{NY} = V_{BY}$.

Hence W_2 reads $V_L I_L \cos(30° - \phi)$ W.

$$\text{Now } (W_1 + W_2) = V_L I_L(\cos 30° \cos\phi - \sin 30° \sin\phi + \cos 30° \cos\phi + \sin 30° \sin\phi)$$

$$= \sqrt{3}\, V_L I_L \cos\phi = \text{three-phase power.}$$

N.B. When $\phi = 60°$, (i.e. $\cos\phi = 0{\cdot}5$ lag), wattmeter W_1 reads zero. If $\cos\phi < 0{\cdot}5$ lag, W_1 reads 'backwards' and the reading is negative.

$$\text{Also} \qquad (W_2 - W_1) = V_L I_L \sin\phi$$

$$\therefore \tan\phi = \sqrt{3}\left\{\frac{W_2 - W_1}{W_2 + W_1}\right\}$$

whence $\cos\phi$ can be determined.

Numerical part: $\quad 20 \text{ h.p.} = 20 \times 746\text{W}.$

$$\text{With efficiency } 0{\cdot}88 \text{ p.u., Input} = \frac{20 \times 746}{0{\cdot}88} = 16\,950\text{W}$$

$$= \sqrt{3}\, V_L I_L \cos\phi$$

$$\therefore \quad I_L = \frac{16\,950}{\sqrt{3} \times 400 \times 0{\cdot}84} = 29{\cdot}3\text{A}.$$

$$\text{Then } W_1 = V_L I_L \cos(30° + \phi)$$

$$= 400 \times 29{\cdot}3 \times 0{\cdot}457 = 5350\text{W}.$$

$$W_2 = V_L I_L \cos(30° - \phi) = 400 \times 29{\cdot}3 \times 1 = 11\,650\text{W}.$$

EXAMPLE 4.4.

A back-to-back efficiency test on two similar d.c. shunt machines gave the following results; supply voltage = 400V, supply current = 36·5A, current from generator = 226·5A, motor field current = 3A, generator field current = 3·5A.
(a) assuming equal copper and core losses, calculate the approximate efficiency of each machine,
(b) if each machine has armature resistance of 0·05Ω, calculate the efficiency of each machine.
Solution:

(a) Using the notation on p. 91 and Fig. 4.11...

$$\eta = \sqrt{\left(\frac{I_2}{I_1 + I_2}\right)} = \sqrt{\left(\frac{226{\cdot}5}{36{\cdot}5 + 226{\cdot}5}\right)} = 0{\cdot}928.$$

(b)

$$\text{Set input} = 400 \times 36{\cdot}5 = 14\,600\text{W}$$

$$\text{Motor } I_a^2 R_a = (260)^2 \times 0{\cdot}05 = 3380$$

$$\text{Generator } I_a^2 R_a = (230)^2 \times 0{\cdot}05 = 2695$$

$$\text{Motor } I_f^2 R_f = V I_{fm} = 400 \times 3 = 1200$$

Generator $VI_{fg} = 400 \times 3{\cdot}5 = 1400.$

$\therefore$ Set total copper losses = 8675W

$\therefore$ Stray losses/machine = $\frac{1}{2}(14\,600 - 8675) = 2962$W.

$$\eta_m = \frac{\text{Input} - \text{losses}}{\text{Input}}$$

$$= \frac{400 \times 263 - (3380 + 1200 + 2962)}{400 \times 263} = 0{\cdot}9283.$$

$$\eta_g = \frac{\text{Output}}{\text{Output} + \text{losses}}$$

$$= \frac{400 \times 226{\cdot}5}{400 \times 226{\cdot}5 + (2695 + 1400 + 2962)} = 0{\cdot}9277.$$

5 D.C. machine performance

Classification of d.c. machines

The method of field excitation determines the name applied to a d.c. machine.

Permanent-magnet machines employ one of the new permanent-magnet materials, e.g. Alnico, to provide the field excitation. These machines are generally of very small power, (0–5 h.p.), and they are used in control systems. They have problems of demagnetization and of stabilization of the magnet system. The permanent-magnet materials are brittle and hard and generally cannot be machined. Compared with machines having electro-magnetic excitation, the p.m. machine has higher efficiency, less weight and no field windings but it is more expensive. It operates at lower flux densities and has a smaller airgap, (Ref. 5).

Most d.c. machines have electro-magnetic excitation and are classified according to the source of excitation current, namely separate-excitation, self-excited shunt, self-excited series and compound, (Fig. 5.1).

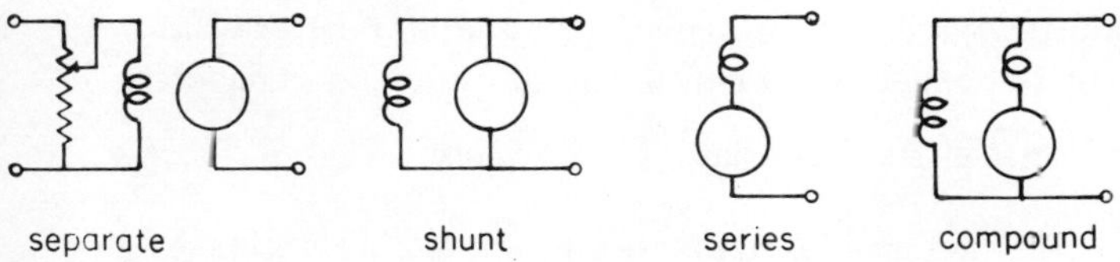

Fig. 5.1.

Examination of steady-state d.c. characteristics provides an understanding of the role played by inherent non-linearities in the operation of the machine.

The magnetization curve

Since generated volts, $e_a = \omega M_d i_f$, the graph of e_a versus i_f at a constant speed, will display proportionally the magnetization characteristics of the machine. To obtain the graph, the machine is operated as a separately-excited generator. Typical characteristics show the measure of residual magnetization, hysteresis, and the incidence of saturation and are shown in Fig. 5.2 for various fixed speeds.

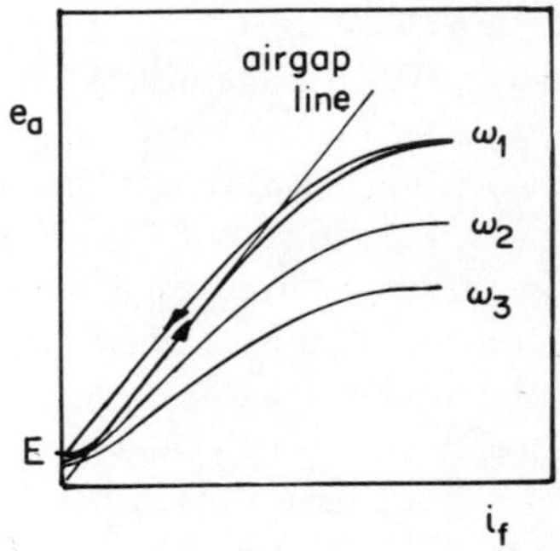

Fig. 5.2.

The initial linear part of the curve extended is termed the Air-gap Line and represents the magnetization curve when saturation in the iron is neglected.

Self-excitation.

Consider a self-excited shunt generator. Due to residual magnetism, a small voltage, E, is generated before the field circuit is connected. When the field circuit is connected:

$$\text{Initial field-current, } i = \frac{E}{R} \text{ where } R = R_f + R_a$$

($R_a \ll R_f$ and may be neglected in these calculations).

$$\text{Initial generated volts} = \omega M_d i$$
$$= ki \text{ where } k = \text{slope of } a\text{–}g \text{ line}$$
and is $\propto$ speed.

$$\text{Then } e_a = ki$$

$$= iR_f + L\frac{di}{dt} \text{ where } L = L_f + L_a$$

$$L\frac{di}{dt} + (R_f - k)i = 0$$

$$\therefore \quad \frac{di}{dt} = \frac{(k - R_f)i}{L}.$$

If $R_f < k$, the field current and e_a will both increase. But because of saturation k is not maintained constant. A stable steady-state value of e_a is reached when $e_a = iR_f$.

This value is obtained where the plot of the R_f intersects the magnetization curve, (Fig. 5.3).

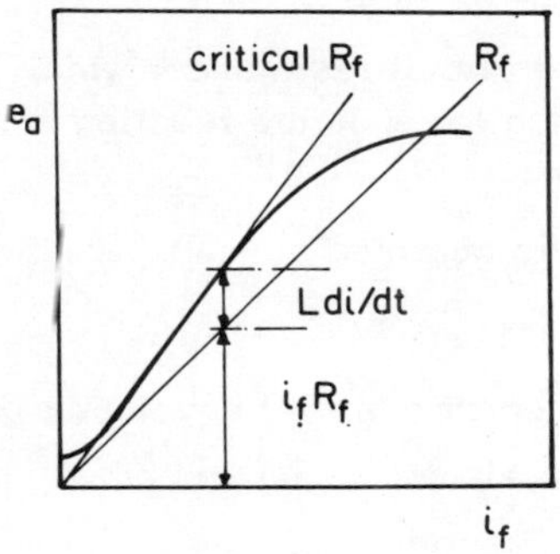

Fig. 5.3.

The natural response equation has a time-constant, $T = L/(k - R_f)$ and thus, as the value of $R_f \to k$, the build up time is longer. For some machines this may be several seconds. The Critical Resistance of the field circuit occurs when $R_f = k$, and will depend on the speed.

Hence the shunt-excited d.c. generator will not build-up a generated voltage:

(a) if there is no initial permanent magnetism,

(b) if the resistance of the field circuit $> k$,

(c) if the field is connected the wrong way round, opposing the permanent magnetism, (i.e. k is negative),

(d) if the driving speed is such as to make $k < R_f$.

For a given R_f, voltage does not build up until the speed has reached the appropriate value or 'Critical speed'.

For the series generator similar principles to those stated above apply, but in this case the field current is also armature current. The characteristics are discussed under 'load characteristics' below.

Steady-state load characteristics

Generator operation.

The load or external characteristic shows the variation in terminal voltage with load current. Generally, constant driving speed is assumed and the relation can be expressed:

$$V = E - I_a R_a$$

where V = terminal voltage,

E = generated voltage,

R_a = armature circuit resistance, which may include an allowance for armature reaction and brush-contact resistance.

A power equation is obtained by multiplying across by I_a, viz.

$$VI_a = EI_a - I_a^2 R_a$$

and EI_a = gross electrical power developed in the armature,

VI_a = armature electrical output,

$I_a^2 R_a$ = armature copper loss.

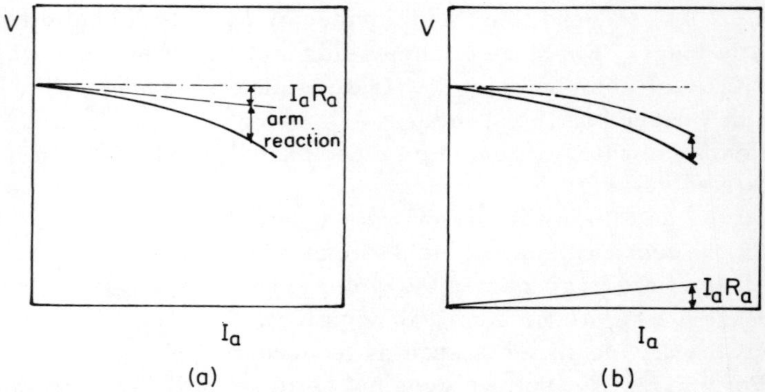

Fig. 5.4.

Separately-excited. Allowing for the ohmic $I_a R_a$ drop, the characteristic curve drops proportionately with load current, (Fig. 5.4(a)). But the experimental curve will drop more steeply due to the de-magnetizing effect of armature reaction.

If the ohmic $I_a R_a$ volt drop is added to the experimental curve, the 'internal' characteristic is obtained, and represents the actual flux conditions in the airgap, Fig. 5.4(b).

Self-excited shunt. In this case:

$$V = E - I_a R_a$$
$$= E - (I_L + I_f) R_a.$$

Note that E is a function of I_f and $I_f = V/R_f$.

Now V decreases due to increasing $I_a R_a$ as in the previous case, but E also decreases due to the fall in I_f. Hence the external characteristic drops more than in the separately-excited case.

As load current is further increased (i.e. R_L reduced), a stage is reached when the fall in E has greater effect on the value of the load current than the decrease in R_L, and I_L begins to decrease, or the characteristic 'turns back'.

This feature may be derived graphically by recognising the component volt drops as determined by the R_f-line and the magnetization curve or open-circuit characteristic, (O.C.C.), Fig. 5.5. As V varies with load, the corresponding ordinates of $I_a R_a$ are determined between the two graphs. Hence, assuming R_a = constant, the typical external characteristic for a shunt generator is derived.

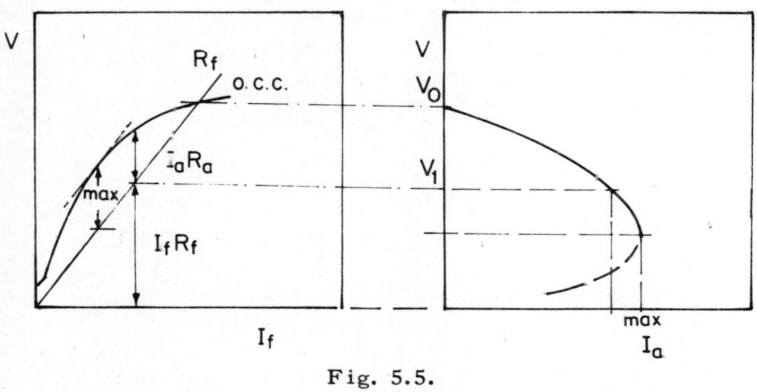

Fig. 5.5.

Self-excited series and compound. For the series case:

$$V = E - I_a (R_a + R_f)$$

$$\text{and} \quad I_f = I_a .$$

Hence the load characteristic is similar in shape to the magnetization curve, Fig. 5.6. As before, allowance for the ohmic drop, $I_a (R_a + R_f)$, determines the Internal characteristic. Note that the latter part of the external characteristic bends over. When worked on the initial straight part, a generated voltage proportional to load current is obtained. On the latter part, if steep enough, an approximately constant current supply is obtained. The 'critical resistance' for self-excitation is determined by the resistance of the external load, (Fig. 5.6).

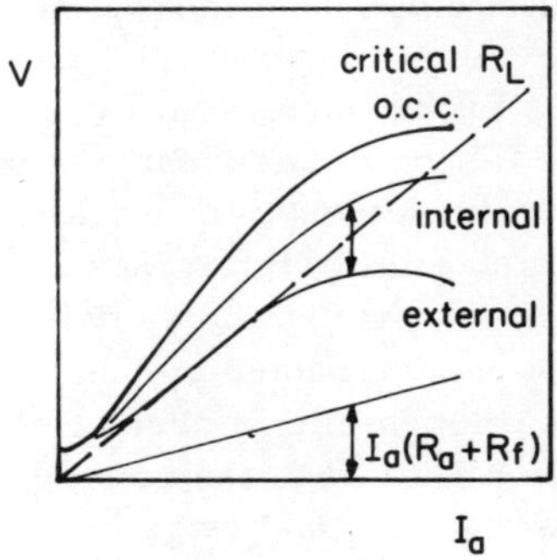

Fig. 5.6.

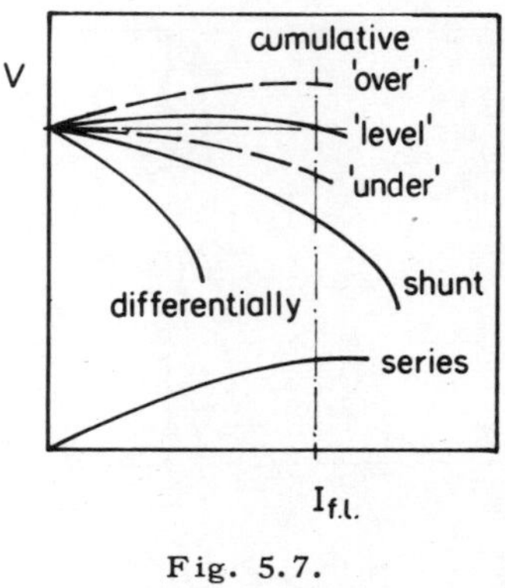

Fig. 5.7.

Series excitation is used as an addition to shunt excitation to provide a boost voltage. The combined characteristics, added, provide a near-constant voltage, Fig. 5.7. The machine is termed a cumulatively compound generator, and in the case cited is 'level' compounded. If the series excitation opposes the shunt, the machine is differentially compounded. This characteristic drops very sharply and the machine has limited application, e.g. electric arcs.

Motor operation.

The d.c. machine is an energy conversion device and operates as a motor or generator depending on the nature of the input energy, i.e. electrical or mechanical. In practice, a machine intended for use as a motor may require a protective enclosure, e.g. drip-proof, fire-proof etc., arising from the inevitable proximity of the machine and its load. Transition from motor to generator operation or vice-versa will be recognised in some of the following discussion.

Shunt or separately excited. These two cases become the same when the supply voltage to armature and field is the same.

For motoring: $V = E + I_a R_a$

where V = armature terminal supply voltage, generally constant,

E = armature generated voltage, or 'back voltage',

$$= \frac{2p\Phi NZ}{60c} = \omega k\Phi.$$

Hence we may write,

$$E = V - I_a R_a$$

and
$$\omega = \frac{V - I_a R_a}{k\Phi}.$$

The value of $I_a R_a$ is usually less than 5% of V, so that

$$\omega = \frac{V}{k\Phi} \text{ approximately.}$$

All speed-control methods involve application of this relation.

The speed characteristic shows a small linear decrease with increasing I_a, but this may be neutralised by a reduction in Φ due to indirect demagnetization, (increased saturation). Fig. 5.8.

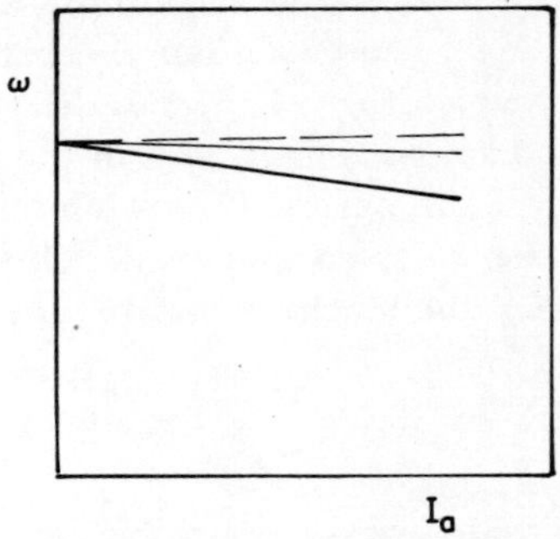

Fig. 5.8.

A power equation is obtained by multiplying the voltage equation by I_a, viz:

$$VI_a = EI_a + I_a^2 R_a$$

where VI_a = electrical power supplied to the armature,

EI_a = gross mechanical power developed by the armature, some of which is absorbed in friction and core losses,

$= \omega Ⓣ$.

Now Torque: $Ⓣ = \dfrac{p\Phi Z I_a}{\pi a} = k\Phi I_a$

and Φ is related to field current. For shunt operation and constant V, Φ is constant. Hence torque varies linearly with armature current. Allowing for frictional torque, nett torque is obtained, Fig. 5.9.

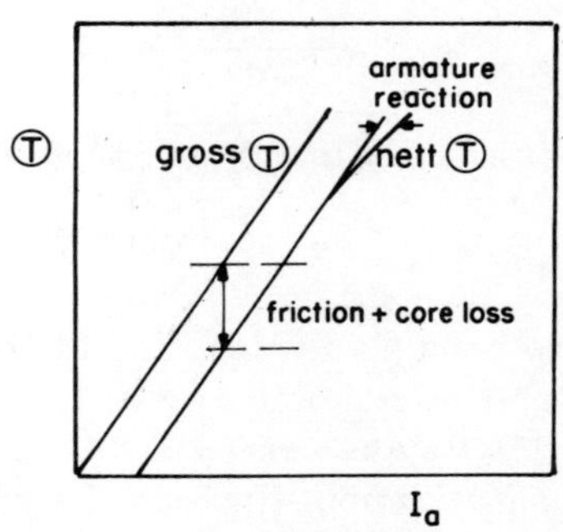

Fig. 5.9.

Although torque is a static effect, it is related to speed since I_a depends on speed, viz:

$$I_a = \frac{V - E}{R_a} = \frac{1}{R_a}(V - \omega k\Phi)$$

$$\therefore \quad \textcircled{T} = \frac{k\Phi}{R_a}(V - \omega k\Phi).$$

If Φ can be assumed constant, $\textcircled{T}$ varies linearly with speed, and maximum (brake) torque occurs when $\omega = 0$, Fig. 5.10. Note that when $V = E$, $\textcircled{T} = 0$, and also $\textcircled{T}$ reverses sign when $E > V$, i.e. in the generating regime. The influence of different fixed field currents is shown.

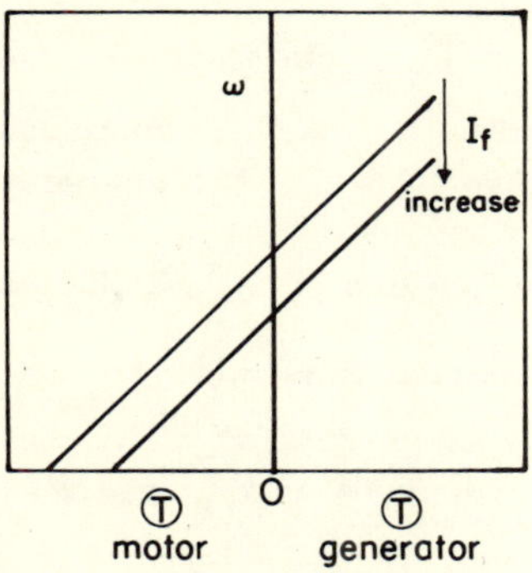

Fig. 5.10.

Series excited. In this case:

$$V = E + I_a R \text{ where } R = (R_a + R_f)$$

$$\text{and} \quad I_f = I_a.$$

Re-arranging the equation as before:

$$E = V - I_a R$$

$$\text{and} \quad \omega = \frac{V - I_a R}{k\Phi}.$$

Neglecting $I_a R$: $\quad \omega = \dfrac{V}{k\Phi}$ approximately.

But $\quad \Phi \propto I_a \quad \therefore \quad \omega \propto \dfrac{1}{I_a}$

and the inverse relation is shown, Fig. 5.11.

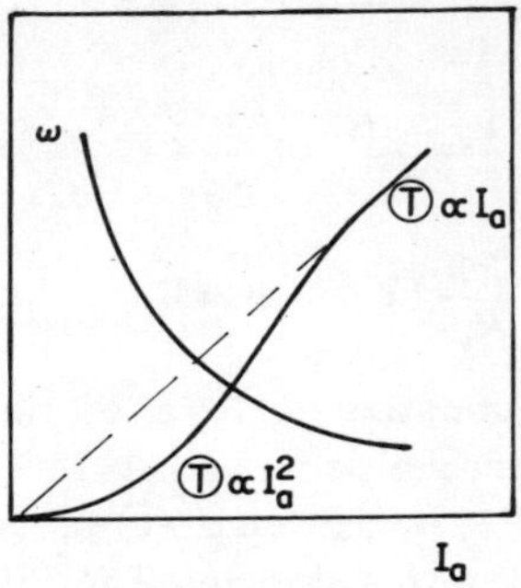

Fig. 5.11.

Now since Ⓣ $= k\Phi I_a$

then Ⓣ $\propto I_a^2$, again since $\Phi \propto I_a$.

This relation is valid until I_f (= I_a) causes magnetic saturation, giving constant Φ, then Ⓣ $\propto I_a$. The characteristic is shown, Fig. 5.11.

Again, torque is a function of I_a, which is a function of speed.

Now $\Phi = \dfrac{V}{k\omega}$ approximately,

then $I_a = \dfrac{V}{K\omega}$ where $\Phi = k_1 I_a$ and $K = kk_1$

$$\therefore \quad Ⓣ = \frac{V^2}{K\omega^2}$$

$$\text{or} \quad \omega \propto \frac{V}{\sqrt{Ⓣ}}$$

Hence the torque/speed characteristic takes the form of Fig. 5.12

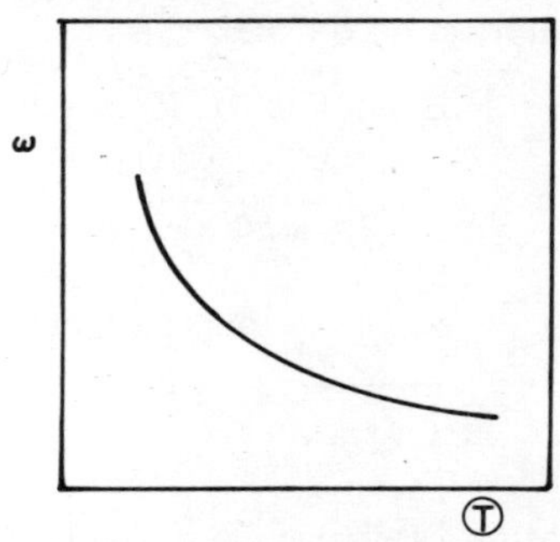

Fig. 5.12.

N.B. Speed becomes dangerously high on light load and hence a series d.c. motor is NEVER run uncoupled to a load.

A popular application is in electric traction since, for a given starting current, Series torque $\propto I^2$, but Shunt torque $\propto I$.

Compound excited. Where series and shunt excitations are combined, the characteristics are a compromise of the two, Fig. 5.13. The presence of shunt characteristics remove the dangers inherent in the series machine, whilst providing better regenerative (braking) characteristics.

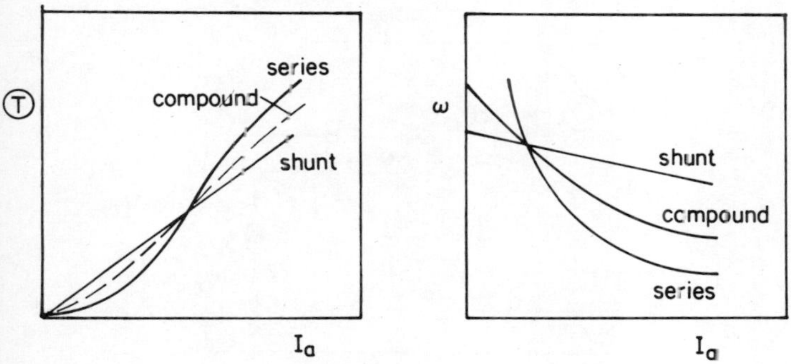

Fig. 5.13.

Starting and speed control

For operation on normal supply voltage, a starting resistance is necessary for all but f.h.p. motors. The starting resistance limits the initial armature current to a safe value until the machine has built up the stable speed and back voltage, E.

If the starting current $= I_1$, (usually 1·5–2 times rated full-load current):

$$\text{Then } R_1 = \frac{\text{supply voltage, } V}{I_1} \text{ where } R_1 = \text{starter resistance} + R_a.$$

As the armature accelerates, the back voltage grows and the current falls to I_2. R_1 is then reduced to R_2, permitting the current to rise to I_1 and the process is continued until the starting resistance is wholly removed and stable running conditions are reached, Fig. 5.14 (a).

For a shunt motor:

Initially: $V = I_1 R_1$ (stud 1).

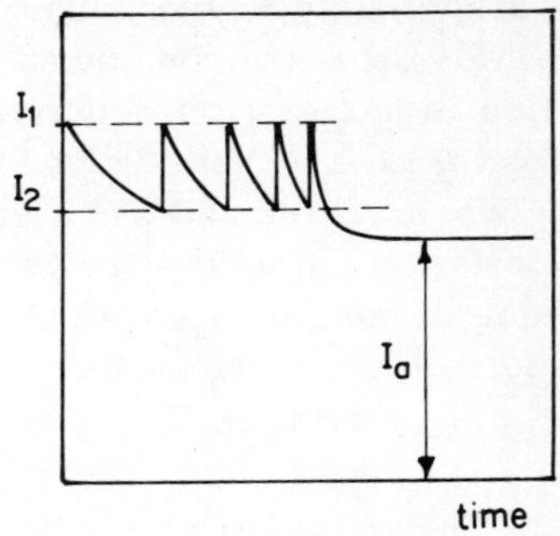

(a)

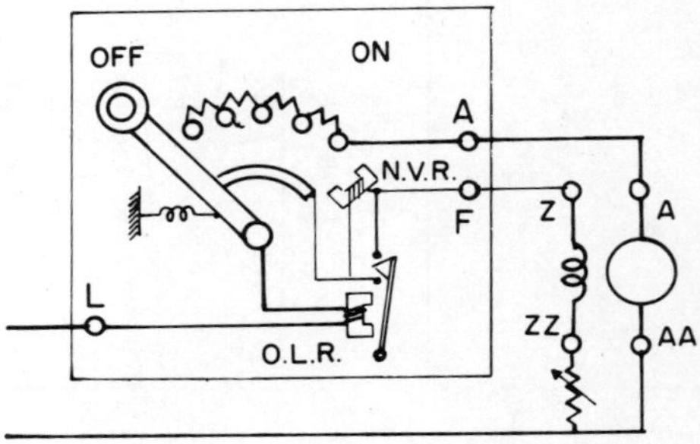

(b)

Fig. 5.14.

At the moment of moving to stud 2, (momentarily constant speed),

$$V = E_1 + I_2 R_1$$
$$= E_1 + I_1 R_2.$$

Repeating the process for further studs, we have:

$$\frac{I_1}{I_2} = \frac{R_1}{R_2} = \frac{R_2}{R_3} = \dots = \frac{R_{n-1}}{R_a} = K$$

where K = the predetermined ratio I_1/I_2 and n = studs on starte

Then $R_1 = KR_2 = K^2 R_3 = \dots = K^{n-1} R_a$

whence n may be determined. Note that the number of resistance sections is $(n - 1)$ and resistance/section $= (R_1 - R_2)$, etc.

Typical features of the hand-operated starter are shown in the diagram, Fig. 5.14 (b). Protective devices include (*i*) an overload trip, (*ii*) no-volt release and hold-on coil, and (*iii*) spring-loaded contact arm to return the starter to 'off' in the event of supply failure etc., and to prevent inadvertent re-connection of the supply voltage without the starter.

For a series motor:

The principle of calculation of the resistance steps is similar, except that, in this case $\Phi \propto I_a$.

Hence at stud 1. $V = I_1 R_1$

at stud 2. $V = E_1 + I_2 R_1 = k\Phi_2 \omega_1 + I_2 R_1$

$= E_1 + I_1 R_2 = k\Phi_1 \omega_1 + I_1 R_2$

at stud 3. $V = E_2 + I_2 R_2 = k\Phi_2 \omega_2 + I_2 R_2$

$= E_2 + I_1 R_3 = k\Phi_1 \omega_2 + I_1 R_3$ etc.

By algebraic manipulation between, say, stud 2 and 3:

$$k\Phi_2 (\omega_2 - \omega_1) = I_2 (R_1 - R_2)$$

$$k\Phi_1 (\omega_2 - \omega_1) = I_1 (R_2 - R_3).$$

$$\therefore \quad \frac{R_1 - R_2}{R_2 - R_3} = \frac{\Phi_2}{\Phi_1} \cdot \frac{I_1}{I_2} = a$$

$$= \frac{r_1}{r_2} = \frac{r_2}{r_3} = \ldots\ldots = \frac{r_{n-1}}{R_a}.$$

If a linear magnetization curve is assumed, $a = 1$ and the resistance steps are equal. Alternately, when I_1 and I_2 are chosen, Φ_1 and Φ_2 are obtained from the magnetization curve. A step-by-step, or a graphical solution is used. Note also:

$$a = \sqrt[n-1]{\frac{r_1}{R_a}}$$

whence n can be checked.

It will be seen that series armature resistance can be used for speed control and thus, in series motors, the starter may also be employed as a speed control. It will then need to be of continuous rating. This type of controller is often used for traction motors.

Time of acceleration:

If I = current for accelerating armature

$=$ mean starter current − load current.

Accel. torque $\text{Ⓣ} = k\Phi I$

$$= J \cdot \frac{d\omega}{dt}$$

$$\therefore \quad \frac{d\omega}{dt} = \frac{Ⓣ}{J} = \frac{k\Phi I}{J}.$$

For constant accel., $\dfrac{d\omega}{dt} = \dfrac{\omega}{t} = \dfrac{Ⓣ}{J}$

$$\text{or,} \quad t = \frac{J\omega}{Ⓣ} = \frac{J\omega}{k\Phi I}.$$

Speed control

For a motor: $\omega = \dfrac{V - I_a R_a}{k\Phi}$.

This equation demonstrates that control of speed is possible by the following methods:

(*i*) *Armature resistance control.*

This method provides a near-linear fall in speed $\propto I_a R_a$. Because of this dependance, speed varies with load and this fluctuation may be undesirable. Control is limited to speeds below normal. The regulating resistance carries I_a and thus the power loss is considerable and the resistance must be continuously rated.

(*ii*) *Field resistance control.*

Speeds above normal are obtained by reducing field current. Excess weakening of the field can produce instability due to the increased influence of armature reaction. But with compensating windings, a speed range up to 5/1 may be obtained. The method is popular and the control resistance is of small current, small power rating.

For the series motor, field control is obtained with a diverter resistance across the field.

Where provision is made for both methods (*i*) and (*ii*), the following consideration can arise:

Since Ⓣ $= k\Phi I_a$, substituting for I_a in the above equation:

$$\therefore \quad \omega = \frac{V}{k\Phi} - \frac{Ⓣ R_a}{k^2\Phi^2}$$

$$\therefore \quad \frac{d\omega}{d\Phi} = -\frac{V}{k\Phi^2} + \frac{2Ⓣ R_a}{k^2\Phi^3} = \frac{2Ⓣ R_a - Vk\Phi}{k^2\Phi^3}.$$

Hence $\dfrac{d\omega}{d\Phi} = 0$ if $2 Ⓣ R_a = Vk\Phi$

$$\text{or} \quad \frac{V}{2} = \frac{Ⓣ R_a}{k\Phi} = I_a R_a .$$

Thus if R_a is such that $I_a R_a = V/2$, a reduction of field current does not mean an increase in speed. An undesirable condition arises and speed may drop with weakened field.

(iii) Controlled supply voltage.

Because $I_a R_a$ is normally a small percentage of V, the speed varies approximately in direct proportion with the voltage. This method is important where a wide range of control is required above and below normal speed. The variable supply voltage may be obtained by controlled rectifiers, or from a separate supply generator which results in the Ward-Leonard system described below, (p. 135). The additional equipment may add considerably to the cost, but this can be justified, particularly where automatic control is involved.

Using silicon controlled rectifiers, (thyristors), the supply to the motor is a chopped d.c. whereby the mean voltage to the machine is varied, Fig. 5.15 (b).

The thyristor, (or SCR), behaves like a transistor rectifier which conducts in the forward direction upon application of a small signal to the 'gate' lead. It is generally triggered from a current source, (or low-impedance voltage source), triggering circuit. A typical SCR characteristic, (Fig. 5.15 (a)) shows that a range of gate currents can cause current avalanche and high conductivity at various values of positive-anode forward-breakdown voltage. Recovery of the 'blocking' state is obtained by, (a) reducing the forward current to a value less than the 'holding' value, 'turn-off' time being of the order of 100 μsec, or (b) reversing the current by application of a reverse voltage between anode and cathode from a low-impedance source, the 'turn-off' time being of the order of 10 μsec.

Typical basic circuit arrangements, (excluding firing circuits), for control of a d.c. machine from a.c. using bridge circuits, or from d.c. mains are shown in Fig. 5.16. The circuits include a 'flywheel' diode, D, which serves the double purpose of dissipating inductive energy of the machine and of prolonging the

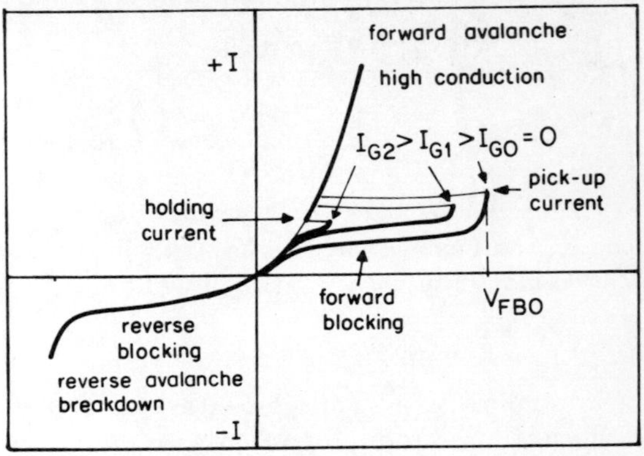

(a)

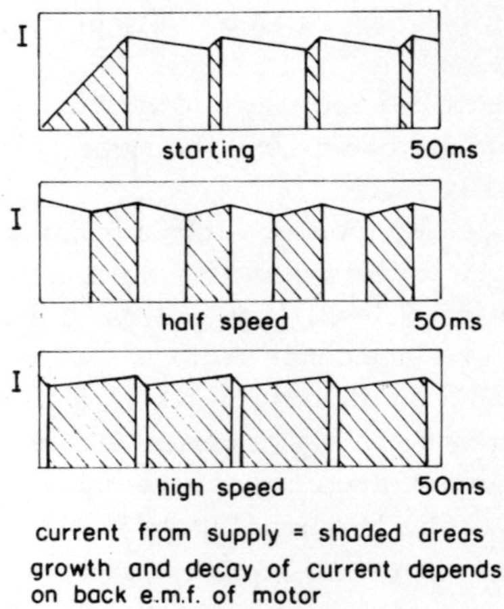

(b)

Fig. 5.15.

flow of armature current during cut-off of the main thyristor, *T*.

Various sophisticated control and trigger arrangements are possible, including means to permit of regenerative braking. Controlled-rectifier methods have particular advantages in applications to d.c. traction motors in respect of maintenance and economic considerations, compared with traditional resistance and contactor methods. (Refs 9, 10).

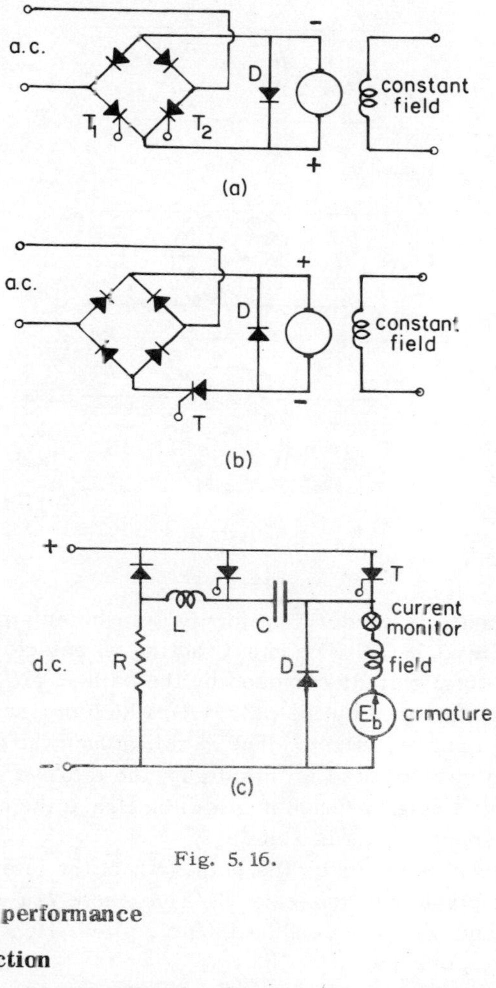

Fig. 5.16.

d.c. machine performance

Armature reaction

A general idea of the effect of armature current, and thereby of armature m.m.f., upon the airgap flux distribution is shown by the elementary two-pole case represented in Fig. 5.17. It is seen that, assuming generator operation, the effect is a weakening of flux under the leading pole tips and a strengthening under trailing pole tips. The flux distortion shows an effective shift of the magnetic neutral axis in the direction of rotation. For motoring action, the effective shift is opposite to the direction of rotation.

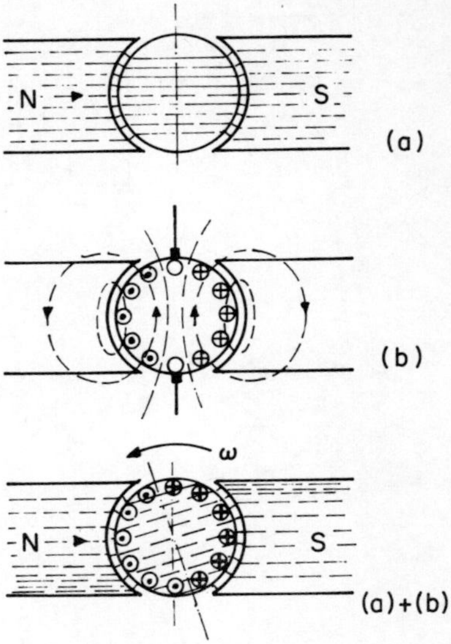

Fig. 5.17.

More exactly, consider a uniformly distributed armature winding, Fig. 5.18 (a). The m.m.f. acting on any closed path equals the total current enclosed by the path = $\oint H \cdot dl$. The nett current enclosed by the path, (A), which crosses the airgap at the pole centres, is zero. The m.m.f. around the closed path is assumed concentrated in the airgap, the relative reluctance of the iron path being assumed negligible. Hence the m.m.f. at each of the pole-centre gaps is zero.

The current enclosed by the path, (B), at the interpolar gaps (i.e. at the brush positions), is $ZI/2pa$ where I/a = current per conductor and $Z/2p$ = conductors/pole pitch. Hence the m.m.f. at each interpolar gap

$$= \tfrac{1}{2}\frac{ZI}{2pa} \text{ amp.turns.}$$

With uniform current distribution, intermediately the m.m.f. varies proportionately, Hence the m.m.f. distribution is triangular, as shown, of peak amplitude

$$= \pm\tfrac{1}{2}\frac{ZI}{2pa}. \text{ (Fig. 5.18 (b)).}$$

Since the airgap is not uniform in this case, a pronounced 'saddle' occurs in the related flux-density distribution, due to the

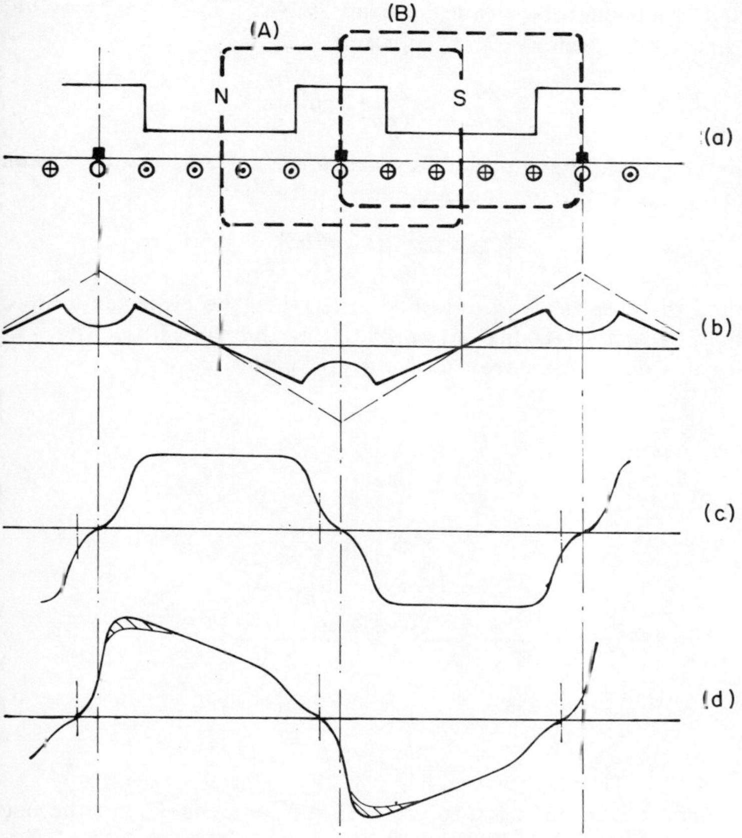

mmf and flux distribution diagrams

Fig. 5.18.

longer airgap. If linearity of magnetization can be assumed, the combined gap flux distribution is obtained by superposition of armature and field flux densities. But due to non-linearity of the magnetization characteristic and, in particular, saturation at pole tips and in armature teeth, the resultant flux density is decreased by the shaded part, (Fig. 5.18 (d)). In other words, the area of the on-load field-form is less than the no-load field-form, and torque and e.m.f. are affected.

Components of reaction.

If the brushes are advanced θ beyond the quadrature-axis, the group of conductors within the band 2θ, (Fig. 5.19) will provide a direct-axis demagnetizing m.m.f.

$$= \frac{2\theta}{\pi}\left(\frac{1}{2} \cdot \frac{ZI}{2pa}\right) \text{amp.turns.}$$

The remaining group provides a cross-magnetizing m.m.f., (torque producing), acting on the quadrature axis, and

$$= \left(1 - \frac{2\theta}{\pi}\right)\left(\frac{1}{2} \cdot \frac{ZI}{2pa}\right).$$

But if the brushes were retarded by θ, then the direct-axis effect would be a magnetizing action assisting the main flux. The reverse conclusions are true for motor operation.

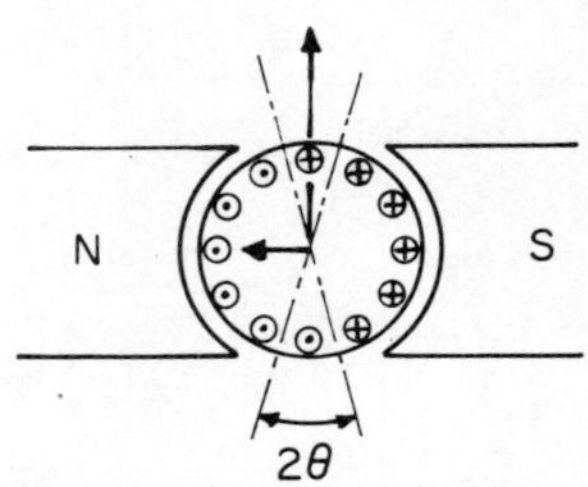

Fig. 5.19.

Hence brush shift, (up to some 15–20° electrical), may be used to compensate on-load loss of flux in a generator, or loss of speed in a motor, consistent of course with satisfactory commutation performance.

Compensating winding.

It has been seen that armature reaction causes distortion of the airgap field-form. Secondly, the voltage induced in an individual conductor is proportional to the local field-form ordinate. Hence an individual coil may have an induced voltage in excess of the average voltage/coil. This can result in flash-over at the commutator segments. The effect can be caused by load fluctuations causing sudden distortion and shifting of the airgap flux. Flux distortion is marked in motors operating on weak fields, (high speeds), at full load or on fluctuating loads.

The problem can be met by placing a compensating winding, in slots in the pole shoes and in series with the armature, to oppose the armature m.m.f. under the pole face and thus reduce peaks of flux distortion. Hence for a compensating winding:

$$\text{Ampere-turns/pole} = \frac{\text{pole arc}}{\text{pole pitch}}\left(\frac{1}{2}\cdot\frac{ZI}{2pa}\right)$$

$$\simeq 0\cdot 7\left(\frac{1}{2}\cdot\frac{ZI}{2pa}\right).$$

Commutation.

Commutation is the switching action between fixed brushes and moving segments, whereby continuous contact is maintained between a brush and the armature winding at the coil which happens to be moving through the magnetic neutral axis. The coil e.m.f. is alternating and the coil current reverses during commutation, thereby maintaining unidirectional current at the brush.

The current reversal is opposed by a 'reactance voltage', $L \cdot di/dt$, where L is the effective inductance of the coil. Inductance L is a composite quantity and depends upon the self-inductance of the coil and mutual inductance of neighbouring coils.

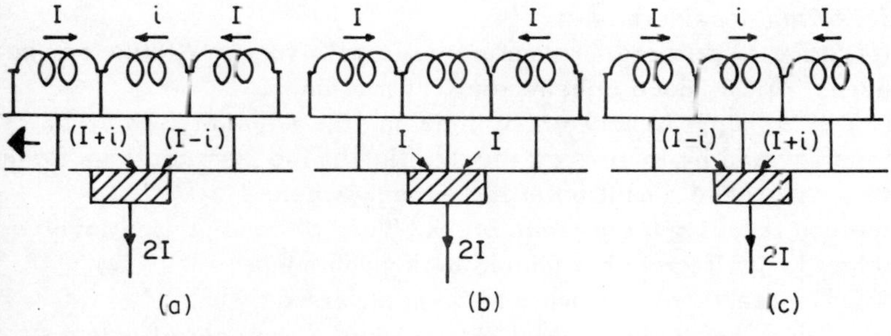

Fig. 5.20.

Time sequence diagrams in Fig. 5.20 show the significant current distributions during the switching action. When $(I - i)$ and R_c, the contact resistance, are both large simultaneously at the diminishing contact, i.e. just before the segment leaves the brush, (Fig. 5.20 (c)), there can be local heating, $(I - i)^2 R_c$, causing ionisation and the formation of an arc. This is undesirable and presents an important problem for the designer. Ideally, linear commutation is sought, (Fig. 5.21), where the desired current reversal occurs in the time of commutation.

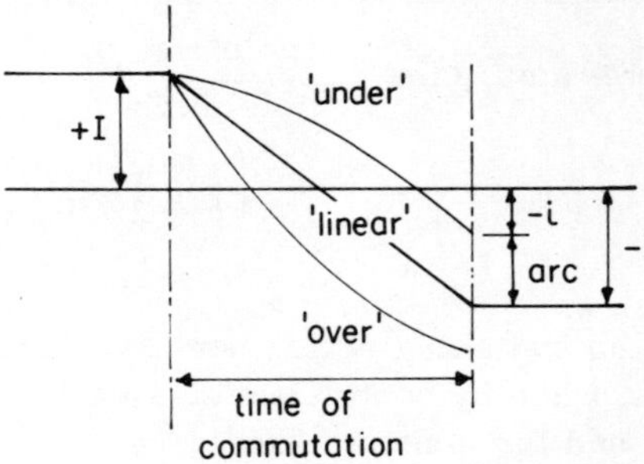

Fig. 5.21.

Methods of improving commutation.

(a) For a given reactance voltage maintaining the arc, by increasing R_c, the arc current $(I - i)$ is reduced and local heating $(I - i)^2 R_c$ will be less. Hence selected grades of carbon brushes, having increased contact resistance, are chosen to improve commutation conditions.

(b) The reactance voltage may be minimized by design from the following considerations:

(*i*) Since $L \propto (\text{turns})^2$, armature coils with very few turns may be used. This is done in heavy-current machines.

(*ii*) di/dt might be reduced by reducing the value of I to be reversed and by increasing the duration of the commutation. Hence we may choose a multipolar lap winding where $I_{\text{coil}} = I/2p$ thereby reducing the current/brush. Time of commutation involves other design factors but should be kept in mind.

(c) The reactance voltage may be neutralised by:

(*i*) advancing the brushes slightly ahead of the neutral axis so that commutation occurs when current reversal in the coil under commutation is already being assisted by the fringing flux of the approaching pole. (N.B. for a motor the brushes are retarded). Since the compensation provided is independent of load current, it is not practical to vary the brush shift with load variations. A compromise setting would be adopted.

(*ii*) providing flux at the neutral axis, of the polarity of the approaching pole, by means of interpoles. The interpole windings are in series with the armature so that the interpole flux is proportional to load current. Hence current reversal in the coil under commutation is assisted for all loadings, including correct polarity interpoles for motor operation.

Interpoles.

In addition to the facts stated in (c) (*ii*) above, interpoles may be seen to perform another important function. Reference to Fig. 5.18 (d) shows that an undesirable flux density may be present on the q-axis due to armature reaction. Interpole m.m.f. should produce neutral zones here, and in addition provide 'commutating flux'. Hence the interpole m.m.f. is some 30% greater than the armature m.m.f. on this axis.

$$\therefore \quad \text{Interpole ampere-turns} = 1{\cdot}3\left(\tfrac{1}{2}\cdot\frac{ZI}{2pa}\right).$$

These narrow auxiliary poles carry the most ampere-turns of any winding in the machine. But being located on the q-axis, they do not affect the total flux/pole on the d-axis. The effect on the airgap flux distribution diagram is shown in Fig. 5.22. To minimize probable saturation at the interpole-core root, the airgap is longer than that of the main pole.

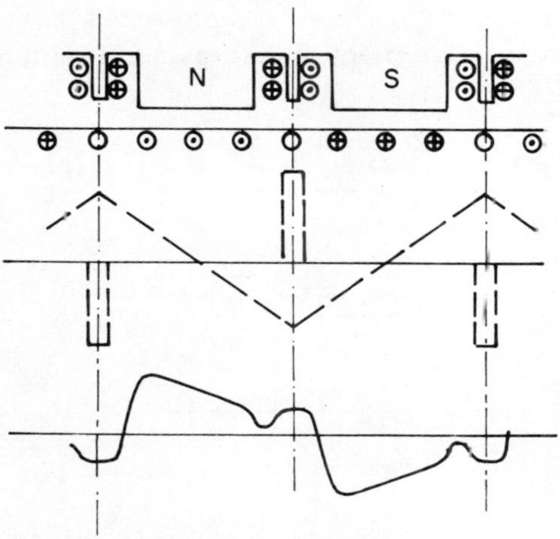

mmf and flux distribution diagrams

Fig. 5.22.

d.c. machines in systems.

The convenience and wide range of control of d.c. machines make them particularly suited for use in control systems. It will also be noted that relatively small power supplied to the field circuit can control a large power in the armature circuit. Hence the machine is a power amplifier.

In practice systems and devices may be subject to periodic forcing functions due to variable loading, pulsating drives or regulating governors. It is important therefore to be able to examine the effects of transient forcing functions and of steady-state periodic forcing functions. Extended study of methods employed in the analysis of systems subjected to these conditions is catered for in textbooks on 'control'. But because of the close relationship with the performance characteristics of devices and machines, some principles are discussed to provide a bridge between the subjects.

Transfer functions

A device or component, which may be part of a system, will affect an input quantity in a manner which will dictate the output quantity. A measure of the character of the device is given by the ratio, output quantity/input quantity. This criterion is termed the Transfer Function of the device. In this way devices, which are of complicated design, can be represented in a manner which conveys the essential effect of their presence in a particular context.

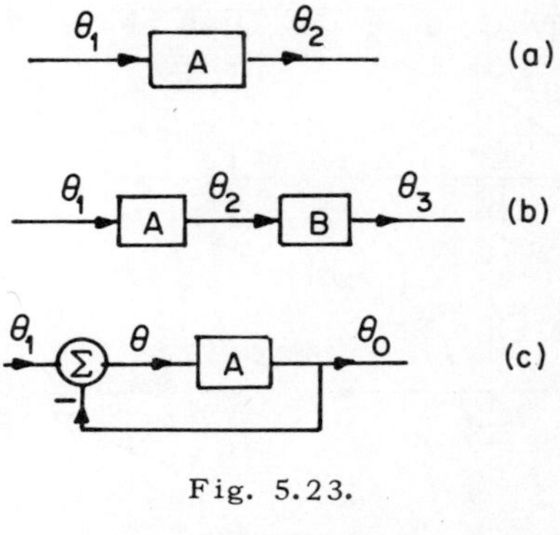

Fig. 5.23.

The transfer function, A, of a device is defined, $A = \theta_2/\theta_1$, and may be represented in a block diagram form, Fig. 5.23(a).

Where a number of devices are arranged in sequence, Fig. 5.23(b):

$$A = \frac{\theta_2}{\theta_1}\ ;\ B = \frac{\theta_3}{\theta_2} \text{ and thus } \frac{\theta_3}{\theta_1} = \frac{\theta_2}{\theta_1} \times \frac{\theta_3}{\theta_2} = A \cdot B$$

and is termed the Overall Transfer Function. This conclusion is valid if A is not affected by the presence of B, i.e. B has infinite input impedance.

Sometimes a feedback connection is made between the output and the input, resulting in a 'closed-loop' system. The closed-loop transfer function is related to the open-loop transfer function as follows; Fig. 5.23(c):

$$\text{Open-loop T.F.} = \frac{\theta_0}{\theta} = A = \frac{\theta_0}{\theta_1 - \theta_0}.$$

$$\text{Closed-loop T.F.} = \frac{\theta_0}{\theta_1} = \frac{1}{1 + \theta/\theta_0} = \frac{1}{1 + 1/A}.$$

These ideas are applied and further developed in the following consideration of the transient performance of d.c. machines.

Linearised equations of the d.c. machine

Using the assumptions of the circuit approach it was shown (p. 66):

$$e_a = \omega M_d i_f.$$

Comparing this with the e.m.f. equation derived from design considerations:

$$M_d = \frac{Z\ \Phi}{\pi a\ i_f} = \frac{\mathrm{d}M}{\mathrm{d}\theta} \text{ at } \theta = \pi/2 \text{ (see p. 66)}.$$

Hence M_d is a constant if Φ/i_f is constant, i.e. saturation is neglected. Also $Z/\pi a$ is the armature conductors linked by the field flux per amp. due to rotation.

The above relationships may be justified from first principles without recourse to dynamic circuit theory as follows:

$$\text{Induced e.m.f. } e = \frac{\mathrm{d}\lambda}{\mathrm{d}t}$$

where λ = mutual armature/field flux linkage.

We may write $\qquad \lambda = M_{af}\ i_f.$

Then $$e = \frac{dM_{af}}{dt} \cdot i_f = \frac{dM_{af}}{d\theta} \cdot \frac{d\theta}{dt} \cdot i_f$$

$$= \omega i_f \frac{dM_{af}}{d\theta}.$$

From Fig. 3.6, $$M_{af} = M_d \cos\theta$$

whence at $\theta = 90°$, $dM_{af}/d\theta = M_d$.

Therefore, $e = \omega M_d i_f$ with brushes on the q-axis. Note also, i_f is d.c. and hence no transformer e.m.f. is involved. (See p.)

Now consider Fig. 5.19, with angle θ designated $d\theta$. Then the differential of armature turns co-axial with the field coil lie within the band $2d\theta$, and

$$dN_{af} = \frac{Z_s}{2\pi} \cdot 2d\theta = \frac{Z}{\pi a} \cdot d\theta$$

$$dM_{af} = \frac{\Phi}{i_f} \cdot dN_{af} = \frac{Z}{\pi a} \cdot \frac{\Phi}{i_f} \cdot d\theta.$$

Then $$\left.\frac{dM_{af}}{d\theta}\right|_{\pi/2}^{0} = \frac{Z}{\pi a} \cdot \frac{\Phi}{i_f} \text{ (Ref. 3).}$$

Expressions for the instantaneous terminal voltages, for motor and generator operation can be stated;*

$$\text{Motor.}\quad v_t = \omega M_d i_f + i_a R_a + L_a p i_a$$

$$\text{Generator.}\quad v_t = \omega M_d i_f - i_a R_a - L_a p i_a .$$

In general the field terminal voltage, (p. 60):

$$v_f = i_f R_f + L_f p i_f + \frac{d}{dt}(M_d i_a).$$

But M_d is assumed constant and also, since the field and armature axes are at 90°, no transformer e.m.f. occurs.

$$\therefore\quad v_f = i_f R_f + L_f p i_f .$$

Equations for torque may be stated as follows:

In general, for a generator:

$$\text{Ⓣ}_s = Jp\omega + D\omega + K_s \int \omega \cdot dt + \text{Ⓣ}_e$$

* Conventions of signs necessary to the Kron theory are introduced later.

where $\textcircled{T}_s$ = mechanical torque at shaft,

$\textcircled{T}_e$ = torque of electrical origin = $M_d i_f i_a$,

D = viscous mechanical friction,

K_s = shaft stiffness, and may be neglected.

Hence for a generator:

$$\textcircled{T}_s = Jp\omega + D\omega + \textcircled{T}_e$$

For a motor:

$$\textcircled{T}_e = Jp\omega + D\omega + \textcircled{T}_s .$$

Transient response of d.c. machines

Principles of dynamic response considered in Chapter 1 are extended and simplified to demonstrate a method of system analysis for the case of one or more interdependant control devices.

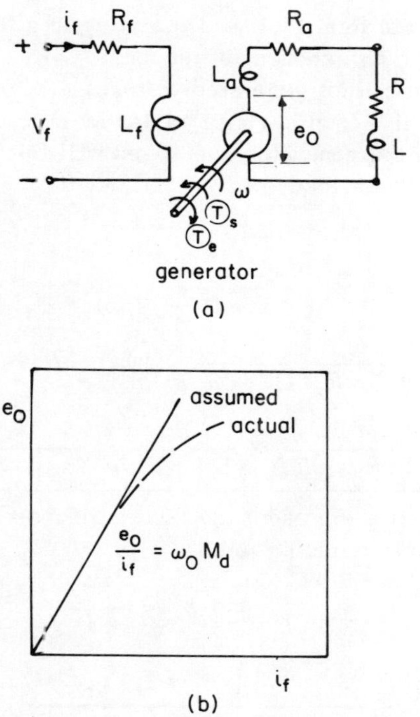

Fig. 5.24.

Consider a d.c. generator, (Fig. 5.24 (a)) driven at constant speed, ω_0. A step-function d.c. voltage, V_f, applied to the field circuit gives the relation:

$$L_f p i_f + i_f R_f = V_f \text{ where } p = \frac{\mathrm{d}}{\mathrm{d}t}.$$

This is a first order differential equation with a solution:

$$i_f = \frac{V_f}{R_f}(1 - \epsilon^{-t/T_f}) \text{ where } T_f = \frac{L_f}{R_f}.$$

Alternately, using the transfer function concept:

$$\frac{i_f}{V_f} = \frac{1}{(R_f + L_f p)} = \frac{1}{R_f} \cdot \frac{1}{(1 + T_f p)}$$

and $$i_f = \frac{V_f}{R_f} \cdot \frac{1}{(1 + T_f p)}.$$

Note: terms of the form $1/(1 + Tp)$ arising in a transfer function are equivalent to an exponential response.

Now the open-circuit generated voltage, $e_0 = \omega_0 M_d i_f$. (Fig. 5.24 (b)). If a load having resistance R and inductance L, is connected to the generator, the speed will fall to ω. Then the generated voltage on load,

$$e_a = \omega/\omega_0 \cdot e_0$$
$$= \frac{\omega M_d}{R_f} \frac{1}{(1 + T_f p)} V_f$$

$$\therefore \quad i_a = \frac{e_a}{(R + R_a)} \frac{1}{(1 + T_a p)} \text{ where } T_a = \frac{L_a + L}{R_a + R}$$

$$= \frac{\omega M_d V_f}{R_f (R + R_a)} \cdot \frac{1}{(1 + T_a p)} \cdot \frac{1}{(1 + T_f p)}.$$

The $(1 + Tp)$ terms represent the factors of a second order differential equation having roots:

$$p_1 = -\frac{1}{T_a} \text{ and } p_2 = -\frac{1}{T_f}$$

whence,

$$i_a = I + A\epsilon^{-t/T_a} + B\epsilon^{-t/T_f}$$

where I = steady-state value and A and B are constants evaluated by substituting conditions at $t = 0$, and $\mathrm{d}i_a/\mathrm{d}t = 0$.

Alternatively, using the time constants and the method of partial fractions, we can express the solution:

$$i_a = I\left\{1 - \frac{T_f}{T_f - T_a}\epsilon^{-t/T_f} - \frac{T_a}{T_a - T_f}\epsilon^{-t/T_a}\right\}.$$

Note on partial fractions:

An expression of the form

$$\frac{1}{(1 + T_1 p)(1 + T_2 p)}$$

can be expressed as

$$\frac{X}{(1 + T_1 p)} + \frac{Y}{(1 + T_2 p)}$$

where $X = \frac{T_1}{(T_1 - T_2)}$ and $Y = \frac{T_2}{(T_2 - T_1)}$.

Substituting for X and Y gives:

$$\frac{T_1}{(T_1 - T_2)}\frac{1}{(1 + T_1 p)} + \frac{T_2}{(T_2 - T_1)}\frac{1}{(1 + T_2 p)}$$

$$= \frac{T_1 T_2 + T_1 T_2^2 p - T_1^2 - T_1^2 T_2 p + T_1 T_2 + T_1^2 T_2 p - T_2^2 - T_1 T_2^2 p}{(T_1 - T_2)(1 + T_1 p)(T_2 - T_1)(1 + T_2 p)}$$

$$= \frac{-T_1^2 + 2T_1 T_2 - T_2^2}{(T_1 - T_2)(T_2 - T_1)(1 + T_1 p)(1 + T_2 p)}$$

$$= \frac{1}{(1 + T_1 p)(1 + T_2 p)}.$$

An approximate solution may be permissible if, for example, the field time constant, T_f, is the principal delay.

Block-diagram representation may be used to organise the mathematical sequence, and can be extended to illustrate the inter-dependance of the several parts of a system.

Thus, the above example may be represented:

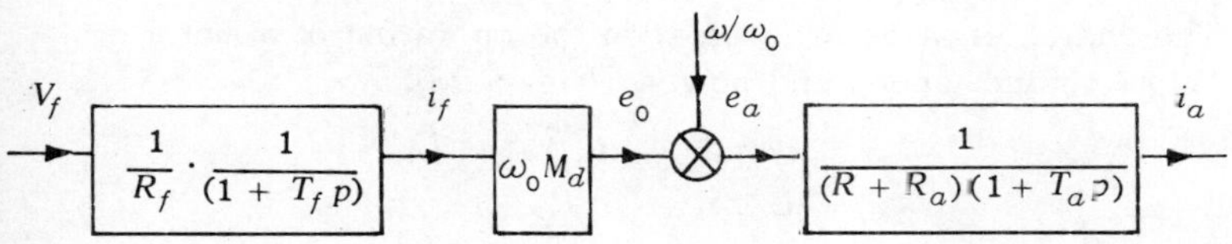

Using the principle of the feed-back loop this diagram may be re-arranged:

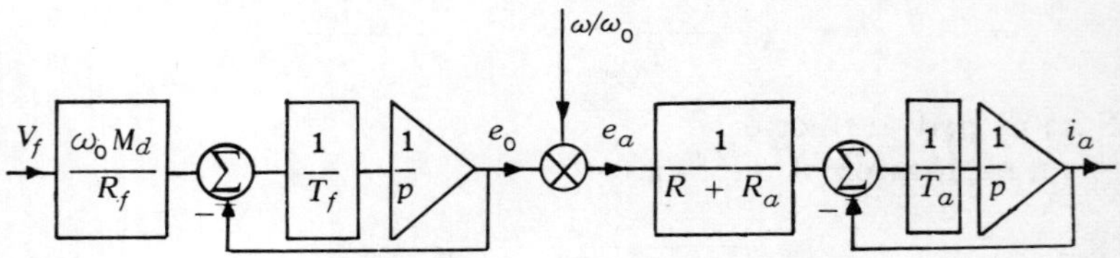

In this form the processes of multiplication, summation, and integration are indicated and the diagram provides guidance for solution of the problem by analogue computer.

The problem and the diagram become more involved as further dependant quantities are added, for example, if consideration of the mechanical drive be added.

Since

$$\textcircled{T}_s = Jp\omega + D\omega + \textcircled{T}_e$$

$$\textcircled{T}_s - \textcircled{T}_e = Jp\omega + D\omega = \omega(Jp + D)$$

$$\therefore \quad \omega = \frac{\textcircled{T}_s - \textcircled{T}_e}{(Jp + D)}.$$

The diagram is re-arranged, in part, as follows:

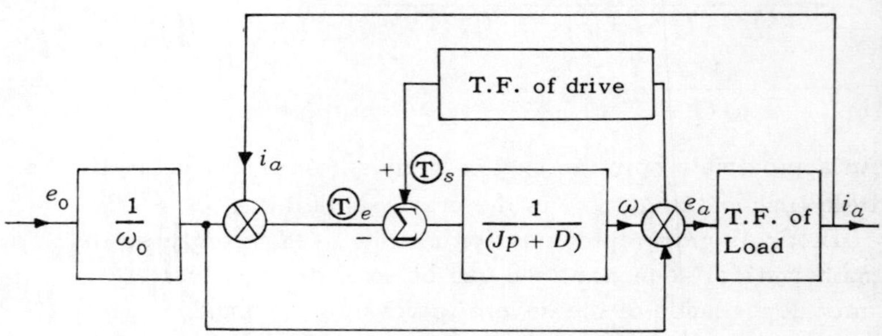

d.c. motors in systems:

The characteristics of interest in the operation of a motor are output torque, speed and power, (Fig. 5.25).

Now

$$v_t = \omega M_d i_f + i_a R_a + L_a p i_a$$

$$v_f = L_f p i_f + i_f R_f$$

$$\textcircled{T}_e = Jp\omega + D\omega + \textcircled{T}_s.$$

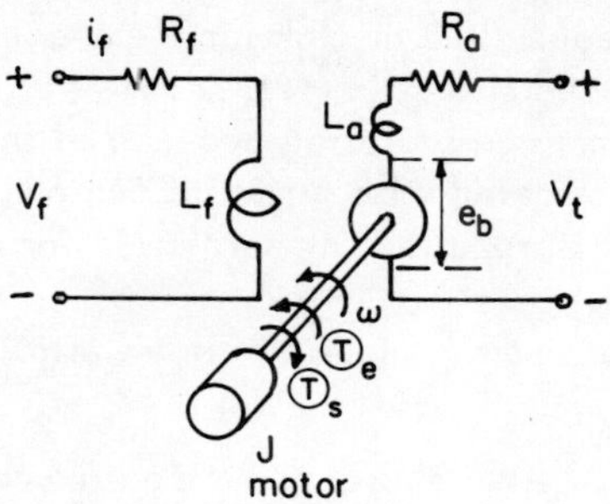

Fig. 5. 25.

(a) consider the case of field control, constant i_a.

As before,

$$i_f = \frac{v_f}{R_f} \cdot \frac{1}{(1 + T_f p)}$$

$$\textcircled{T}_e = M_d i_f i_a = K_a i_f \text{ where } K_a = M_d i_a$$

$$\therefore \textcircled{T}_e = \frac{K_a v_f}{R_f} \cdot \frac{1}{(1 + T_f p)}$$

$$\omega = \frac{\textcircled{T}_e - \textcircled{T}_s}{Jp + D} = \frac{\textcircled{T}_e - \textcircled{T}_s}{D} \cdot \frac{1}{(1 + T_m p)}$$

$$\therefore \omega = \left\{ \frac{K_a v_f}{DR_f} \cdot \frac{1}{(1 + T_f p)} - \frac{\textcircled{T}_s}{D} \cdot \right\} \frac{1}{(1 + T_m p)}$$

where T_m = mechanical time constant = J/D.

We may write :

$$\omega = \left\{ \frac{K_a}{DR_f} \cdot \frac{1}{(1 + T_f p)} \cdot \frac{1}{(1 + T_m p)} \right\} v_f - \left\{ \frac{1}{D(1 - T_n p)} \right\} \textcircled{T}_s$$

to indicate the effect of variable v_f, or variable $\textcircled{T}_s$ upon the speed.

Block diagram representation of this system may be shown :

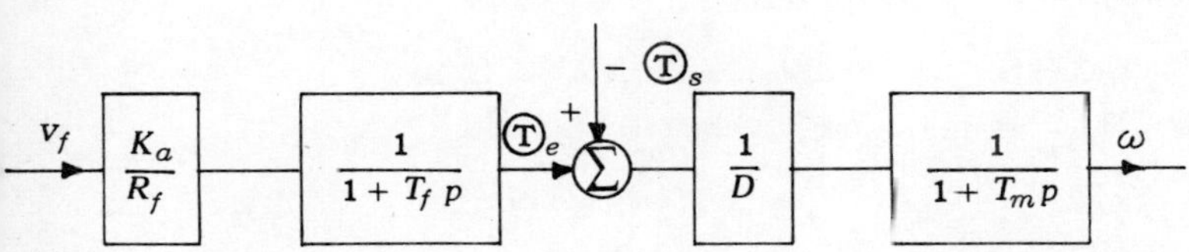

N.B. With constant i_a the mechanical time-constant is determined by mechanical properties only, (cf. p. 14).

Field control may be preferred because of the small control power required compared with armature control. If direction-sensitive control is required, a 'split-field' or centre-tapped field may be employed.

The field is commonly supplied via an amplifier of gain A. Then we have:

$$v_f = Av_c \text{ where } v_c = \text{control volts.}$$

Also, torque becomes:

$$Ⓣ_{eA} = \frac{K_a A v_c}{(R_f + R)} \cdot \frac{1}{(1 + T_{fA}\, p)}$$

where R = internal resistance of the amplifier, and the field-circuit time-constant becomes

$$Ⓣ_{fA} = \frac{L_f}{(R_f + R)}.$$

Hence a high resistance amplifier reduces the effective field time-constant.

Addition of velocity feedback by means of a tachogenerator gives effective control voltage $= (v_c - \omega K_t)$, where K_t = volts/radian/sec for the tachogenerator. The implications of this feedback loop are developed below in connection with the Ward-Leonard system, and in connection with stability (p. 144).

Without feedback the new value for output speed, for the present case, becomes:

$$\omega = \left\{\frac{K_a A}{D(R_f + R)} \; \frac{1}{(1 + T_{fA}\, p)(1 + T_m p)}\right\} v_c - \left\{\frac{1}{D(1 + T_m p)}\right\} Ⓣ_s$$

(see also Example 5.11 below).

(b) armature control, constant i_f.

We have, $v_t = \omega M_d i_f + i_a R_a + L_a p i_a$

L_a may be neglected,

$$\therefore\; v_t = \omega K_f + i_a R_a \text{ where } K_f = M_d i_f.$$

Then $$i_a = \frac{v_t - \omega K_f}{R_a}$$

$$\textcircled{T}_e = M_d i_f i_a = K_f i_a$$
$$= \frac{K_f v_t - \omega K_f^2}{R_a}.$$

But $$\textcircled{T}_e = Jp\omega + D\omega + \textcircled{T}_s$$

$$\frac{K_f v_t - \omega K_f^2}{R_a} = Jp\omega + D\omega + \textcircled{T}_s$$

$$\frac{K_f v_t}{R_a} = Jp\omega + \left(D + \frac{K_f^2}{R_a}\right)\omega + \textcircled{T}_s$$
$$= Jp\omega + \mathrm{D_e}\,\omega + \textcircled{T}_s$$

where $\mathrm{D_e}$ = effective mech. friction and e.m. damping.

$$\omega = \left(\frac{K_f v_t}{R_a} - \textcircled{T}_s\right)\left\{\frac{1}{Jp + \mathrm{D_e}}\right\}$$
$$= \left(\frac{K_f v_t}{\mathrm{D_e} R_c} - \frac{\textcircled{T}_s}{\mathrm{D_e}}\right)\left\{\frac{1}{1 + T_m p}\right\}$$

This may be written :

$$\omega = \left\{\frac{K_f}{\mathrm{D_e} R_a} \cdot \frac{1}{1 + T_m p}\right\} v_t - \left\{\frac{1}{\mathrm{D_e}(1 + T_m p)}\right\}\textcircled{T}_s$$

where T_m = mech. time constant $= J/\mathrm{D_e}$.

When D is negligible $$T_m = \frac{JR_a}{K_f^2}$$

and $$\mathrm{D_e} = \frac{K_f^2}{R_a}.$$

(c) Consider a step-function d.c. voltage, V_t applied to the armature, i_f constant and let $\textcircled{T}_s = 0$.

Neglecting mechanical friction, $\mathrm{D_e} = \dfrac{K_f^2}{R_a}$ and $T_m = \dfrac{JR_a}{K_f^2}$.

Then $$\omega = \frac{K_f V_t}{\mathrm{D_e} R_a}\left(\frac{1}{1 + T_m p}\right)$$

$$= \frac{V_t}{K_f}\left(1 - \epsilon^{-t/T_m}\right)$$

$$i_a = \frac{V_t}{R_a} - \frac{\omega K_f}{R_a}$$

$$= \frac{V_t}{R_a} - \frac{K_f}{R_a}\cdot\frac{V_t}{K_f}\left(1 - \epsilon^{-t/T_m}\right)$$

$$= \frac{V_t}{R_a}\,\epsilon^{-t/T_m}.$$

Now energy from supply during running-up:

$$= V_t \int_0^\infty i_a \cdot \mathrm{d}t$$

$$= V_t \cdot \frac{V_t}{R_a}\int_0^\infty \epsilon^{-t/T_m}\cdot \mathrm{d}t$$

$$= \frac{V_t^2}{R_a}\cdot\frac{JR_a}{K_f^2} = \frac{JV_t^2}{K_f^2}.$$

Steady-state speed, $\omega_{ss} = \dfrac{V_t}{K_f}$

$\therefore$ Energy stored in mass $= \dfrac{1}{2}J\omega_{ss}^2$

$$= \frac{1}{2}\,\frac{JV_t^2}{K_f^2}.$$

Energy dissipated $= \displaystyle\int_0^\infty i_a^2 R_a \cdot \mathrm{d}t$

$$= \frac{V_t^2}{R_a}\int_0^\infty \epsilon^{-2t/T_m}\cdot \mathrm{d}t$$

$$= \frac{V_t^2}{R_a}\cdot\frac{JR_a}{2K_f^2}$$

$$= \frac{1}{2}\,\frac{JV_t^2}{K_f^2}.$$

$\therefore$ Energy supplied = Energy stored + Energy dissipated.

The Ward Leonard system

A popular example of d.c. machines used in a control system is the Ward Leonard method of speed control, Fig. 5.26. In elementary form it consists of a separately-excited d.c. generator, driven at constant speed, which provides the armature input to a constant-field d.c. motor. Control of the motor speed over a wide range is obtained by varying the generator field current. A block diagram can be drawn for the system, Fig. 5.27, based on the cases already considered.

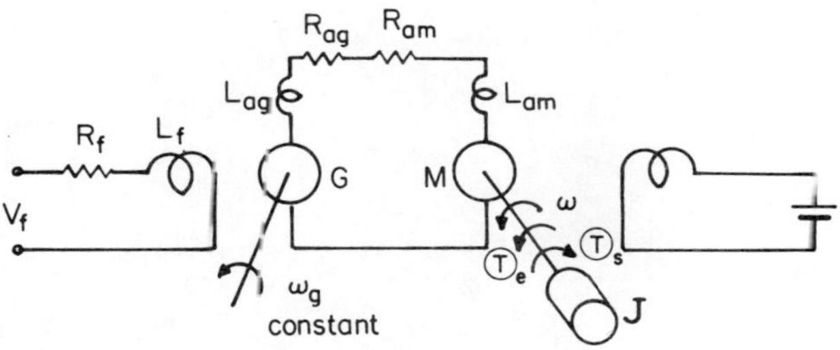

Fig. 5.26.

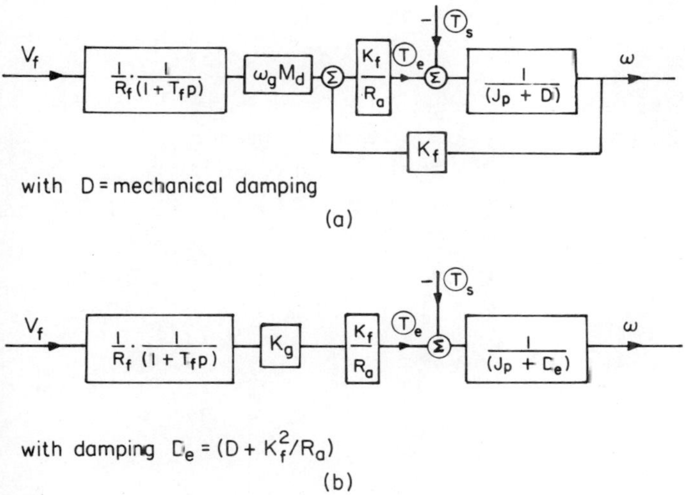

Fig. 5.27.

For the generator part:

$$i_f = \frac{1}{R_f\,(1 + T_f\,p)}\, v_f$$

$$e_g = \omega_g M_d\, i_f$$

$$= \frac{K_g}{R_f\,(1 + T_f\,p)}\, v_f$$

where K_g = slope of the e_g/i_f graph at constant ω_g.

For the motor, neglecting armature self-inductances:

Let K_f = torque/unit arm. current, $R_A = R_{ag} + R_{am}$,

D_e = mechanical and e.m. damping and $J = J_{motor} + J_{load}$.

Then
$$i_a = \frac{e_g - \omega K_f}{R_A}$$

$$Ⓣ_e = K_f\, i_a$$

$$= \frac{K_f\, e_g - \omega K_f^2}{R_A}.$$

But
$$Ⓣ_e = Jp\omega + D_e\,\omega + Ⓣ_s$$

$$\therefore \quad \omega = \left(\frac{K_f\, e_g}{R_A} - Ⓣ_s\right)\left(\frac{1}{Jp + D_e}\right)$$

$$= \frac{1}{D_e}\left(\frac{K_f\, e_g}{R_A} - Ⓣ_s\right)\left(\frac{1}{1 + T_m\, p}\right).$$

Some simplification occurs if mechanical damping can be neglected, and $D_e = K_f^2/R_A$ with a pure inertia load.

For a step input V_f, starting from rest and $Ⓣ_s = 0$

$$\omega = \frac{K_f\, e_g}{D_e\, R_A}\,\frac{1}{(1 + T_m p)}$$

$$= \frac{K_f\, K_g}{D_e\, R_A R_f}\,\frac{1}{(1 + T_f\, p)\,(1 + T_m p)}\, V_f$$

$$= \frac{K_1}{(1 + T_f\, p)\,(1 + T_m p)}\, V_f \text{ radians/sec.}$$

where
$$K_1 = \frac{K_f\, K_g}{D_e\, R_A\, R_f}.$$

Commonly in speed control systems the output speed is monitored, by using a tachometer-generator feed-back loop. The nett discrepancy or error between the command input and the output speed is fed as an error voltage to an amplifier, which, in this case, supplies the generator field voltage, Fig. 5.28.

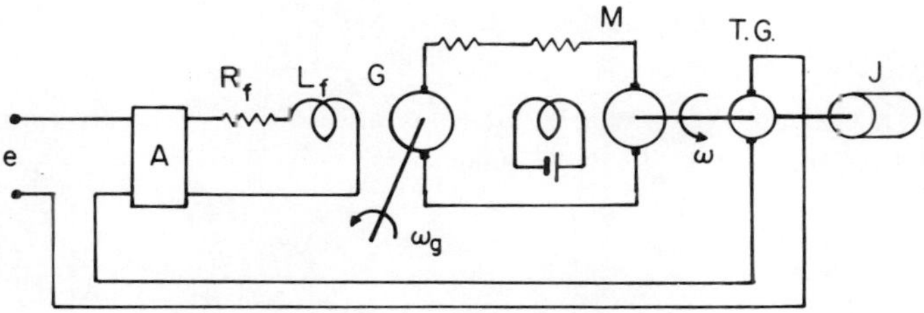

Fig. 5.28.

Let e = command control voltage

and ωK_t = tachometer feedback voltage,

$\therefore v_f = A(e - \omega K_t)$.

Hence $$\omega = \frac{K_1}{(1 + T_f p)(1 + T_m p)} \cdot A(e - \omega K_t)$$

Re-arranging we get:

$$\omega\left[1 + \frac{AK_1 K_t}{(1 + T_f p)(1 + T_m p)}\right] = \frac{K_1 Ae}{(1 + T_f p)(1 + T_m p)}$$

$$\omega = \left[\frac{K_1 A}{(1 + T_f p)(1 + T_m p) + AK_1 K_t}\right] e$$

$$= \frac{K}{K_t}\left[\frac{1}{(1 + T_f p)(1 + T_m p) + K}\right] e$$

where $K = AK_1 K_t$.

The expression may be re-written:

$$\omega = \frac{1}{K_t}\left(\frac{K}{K + 1}\right)\left[\frac{1}{p^2\left(\frac{T_f T_m}{K + 1}\right) + p\left(\frac{T_f + T_m}{K + 1}\right) + 1}\right] e.$$

The transfer function of the system, with p replaced by complex frequency s, provides us with the second order equation, the roots of which determine the natural frequencies of the system. Comparing with the general solution, (pp. 10, 14):

$$\text{Undamped natural frequency } \omega_n = \sqrt{\left(\frac{1 + K}{T_f\, T_m}\right)}$$

$$\text{and Damping Factor } a = \frac{1}{2}\left(\frac{1}{T_f} + \frac{1}{T_m}\right).$$

When the motor is initially at rest, and $a < \omega_n$, the response for a step input $e = E$ is given:

$$\omega = \frac{E}{K_t}\left(\frac{K}{1 + K}\right)\left[1 - \frac{\omega_n}{\omega_d}\,\epsilon^{-at}\cos(\omega_d t - \phi)\right]$$

where $\omega_d = \omega_n(1 - \zeta^2)^{\frac{1}{2}}$ and Damping Ratio, $\zeta = a/\omega_n$. Hence the transient response is a damped sinusoidal oscillation, Fig. 5.29.

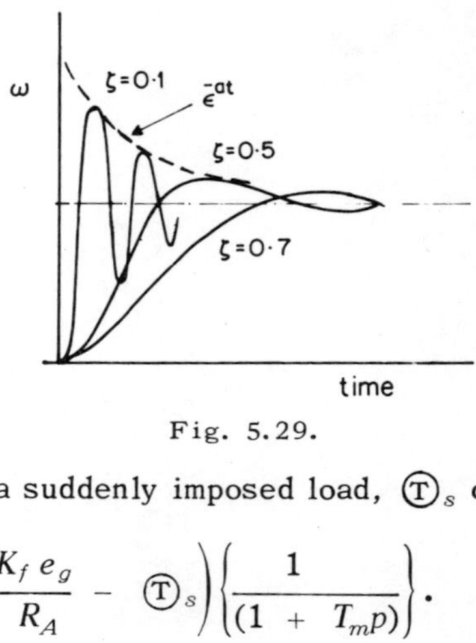

Fig. 5.29.

The effect of a suddenly imposed load, Ⓣ$_s$ can be seen as follows:

$$\text{Since } \omega = \frac{1}{D_e}\left(\frac{K_f\, e_g}{R_A} - Ⓣ_s\right)\left\{\frac{1}{(1 + T_m p)}\right\}.$$

Then with $e = 0$, but with the tachogenerator feedback loop:

$$\omega = \frac{1}{D_e}\left(\left\{\frac{-K_f\,\omega A K_t\, K_g}{R_A R_f (1 + T_f\, p)}\right\} - Ⓣ_s\left\{\frac{1}{(1 + T_m p)}\right\}\right).$$

Substituting K_1 and re-arranging:

$$\omega\left\{1 + \frac{AK_1 K_t}{(1 + T_f\,p)(1 + T_m p)}\right\} = \frac{-\text{Ⓣ}_s}{D_e (1 + T_m p)}$$

$$\therefore \quad \omega = -\frac{(1 + T_f\,p)}{(1 + T_f\,p)(1 + T_m p) + AK_1 K_t} \cdot \frac{T_s}{D_e}.$$

Resolution of the denominator, as before, shows that a similar damped sinusoidal response occurs. Under steady-state:

$$\frac{\Delta\omega}{\Delta \text{Ⓣ}_s} = \frac{R_A}{K_f^2}\left(\frac{1}{1 + K}\right) \text{assuming } D_e = \frac{K_f^2}{R_A}.$$

or,

$$\frac{\Delta\omega}{\Delta \text{Ⓣ}_s} = \frac{R_A}{K_f^2}\left(\frac{1}{1 + AK_1 K_t}\right).$$

Hence if the system is to be insensitive to load disturbances, R_A should be small and K_f large. Also K should be large, suggesting higher amplifier gain. But increased gain increases ω_n and decreases ζ. In practice a damping ratio between 0·4 and 0·7 is satisfactory. Note that a, which decides the duration of the transient disturbance, is determined by the machine time constants. N.B. R_f and T_f include amplifier resistance.

Crossfield d.c. machines

This is an important class of d.c. machines commonly employed in control systems. Notable characteristics are the faster response times and greater amplification obtained compared with ordinary d.c. machines.

The metadyne.

Consider a normal d.c. generator, Fig. 5.30 (a) with brushes located on the q-axis. The generated voltage depends on the main field, Φ_d and speed of rotation. Then:

$$e_{q0} = \omega G_{qf}\, i_f$$

where G_{qf} is the rotational inductance coefficient $= dM/d\theta = M_d$ with brushes on the q-axis.

If the q-axis brushes are short-circuited, a heavy current flows in the armature conductors and this produces a strong field Φ_q, directed along the q-axis. Fig. 5.30 (b). Additional brushes are now located on the d-axis and a d-axis generated voltage is obtained, $e_d = \omega G_{dq} i_q$, Fig. 5.30 (c).

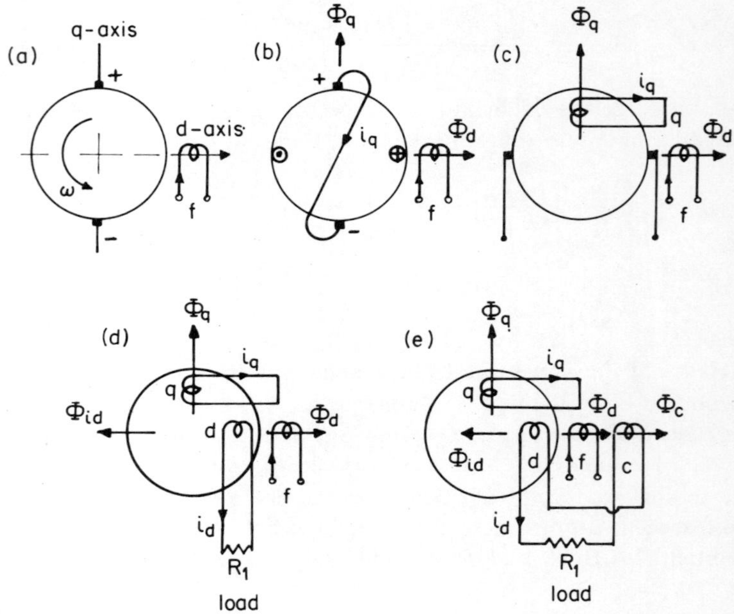

Fig. 5.30.

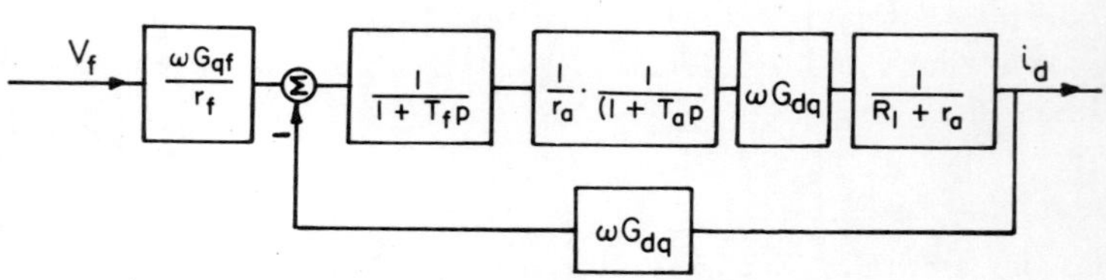

Fig. 5.31.

Division of the armature winding by the brush pairs may be represented schematically by simple d and q coils, Fig. 5.30 (d). Now, current i_d in the d-coil due to a load connected at the d-brushes, will produce a flux Φ_{id} and this opposes the main field, Fig. 5.30 (d). This machine is termed a 'metadyne' and is effectively a two-stage negative-current-feedback-amplifier with a very short time constant.

A block diagram may be drawn to define the interdependance

of the stages, Fig. 5.31.

Equations for the first stage are therefore:

$$e_q = \frac{(\omega G_{qf} V_f / r_f - \omega G_{qd} i_d)}{(1 + T_f p)}$$

$$i_q = \frac{e_q}{r_a} \frac{1}{(1 + T_a p)}.$$

Equations for the second stage are:

$$e_d = \omega G_{dq} i_q$$

$$i_d = \frac{e_d}{R_1 + r_a}.$$

For sinusoidal flux distribution and uniform airgap:

$$G_{dq} = - G_{qd} = L_d = L_q.$$

Under steady-state conditions:

$$I_d = \frac{\omega G_{dq} (\omega G_{qf} I_f - \omega G_{qd} I_d)}{(R_1 + r_a) r_a}$$ which may be stated

$$= \frac{k_1 (k_2 I_f - k_3 I_d)}{(R_1 + r_a) r_a}$$

whence

$$I_d = \frac{k_1 k_2}{r_a (R_1 + r_a) + k_1 k_3} I_f .$$

Since $k_1 k_3 \gg r_a (R_1 + r_a)$, the Metadyne behaves as a current source, with an output current proportional to field current.

The amplidyne.

The power amplification can be improved by elimination of the negative feedback. This is achieved by the introduction of a 'compensating winding' on the main field axis, connected in series with the load, to oppose the flux produced by the *d*-axis rotor current. Fig. 5.30 (e). With exact compensation the block diagram becomes, Fig. 5.32.

The terminal voltage at the load is then:

$$i_d R_1 = \frac{R_1}{r_a + R_1} \cdot \frac{K}{r_f} \cdot \frac{1}{(1 + T_f p)(1 + T_a p)} V_f$$

where $$K = \omega^2 G_{qf} G_{dq}.$$

Under steady-state conditions:

$$\text{Power input} = \frac{V_f^2}{r_f}$$

$$\text{Power output} = \frac{V_f^2 K^2 R_1}{(r_a + R_1)^2}$$

$$\text{Power amplification} \doteq \frac{K^2 r_f}{R_1} \text{ when } R_1 \gg r_a.$$

Power amplification of the order of 10,000 can be obtained compared with 20 to 200 with a single ordinary d.c. generator. Note that the amplidyne is a voltage source. The time constant is not as low as for the metadyne.

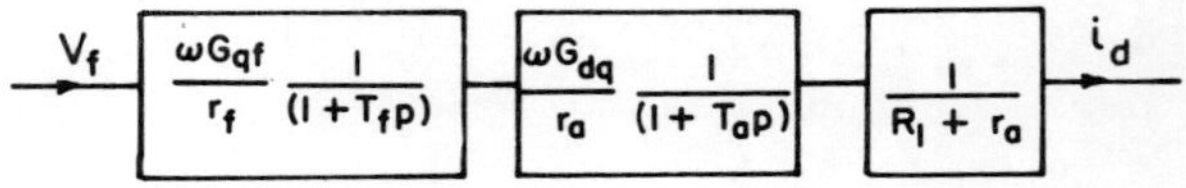

Fig. 5.32.

Frequency response and stability

Study of the operation of a system over a suitable range of frequency is termed steady-state frequency response analysis. This can provide valuable information regarding the stability of a projected system when examined mathematically. Since the frequency-response transfer function is a complex quantity, it is examined in terms of magnitude and phase. Plotted as a vector locus it is termed a Nyquist diagram. Plotted on a frequency axis it is termed a Bode plot.

Some principles are demonstrated for the case of a d.c. machine used in a position-control system. Consider the case of a constant-field motor with armature control. In terms of shaft position, θ:

$$v_t = i_a R_a + K_f \cdot \frac{d\theta}{dt} \quad \text{where } K_f = M_d i_f$$

and $d\theta/dt$ = rotational speed

$$\text{Ⓣ}_e = M_d i_f i_a$$

$$= K_f i_a.$$

Also $Ⓣ_e = Jp^2\theta + Dp\theta$

$$\therefore \quad i_a = \frac{J}{K_f} \cdot p^2\theta + \frac{D}{K_f} \cdot p\theta.$$

ʖubstituting in equation above:

$$v_t = \frac{JR_a}{K_f} \cdot p^2\theta + \frac{DR_a}{K_f} \cdot p\theta + K_f \cdot p\theta$$

$$\frac{v_t}{K_f} = \frac{JR_a}{K_f^2} \cdot p^2\theta + \left(\frac{DR_a}{K_f^2} + 1\right) \cdot p\theta.$$

ʿhis may be expressed:

$$\frac{v_t}{K_f} = T_a \cdot p^2\theta + K_1 \cdot p\theta \text{ where } T_a = \frac{JR_a}{K_f^2}$$

$$\text{and} \quad K_1 = \left(\frac{DR_a}{K_f^2} + 1\right).$$

ʿhen $$\frac{\theta}{v_t} = \frac{1}{K_1 K_f p\left(1 + \frac{T_a}{K_1} \cdot p\right)}$$

$$= \frac{1}{Kp(1 + Tp)} \text{ where } T = \frac{T_a}{K_1} \text{ and } K = K_1 K_f.$$

The steady-state frequency response is obtained by substituting ω for the p-operator:

$$\therefore \quad \frac{\theta}{v_t} = \frac{1}{j\omega K(1 + j\omega T)}.$$

This transfer function is the product of

$$-\frac{j}{\omega K} \text{ and } \frac{1}{(1 + j\omega T)},$$

ɲd these may be seen as vector loci, (Fig. 5.33). The product of ectors involves the multiplication of the magnitudes and the ddition of the angles. Hence the frequency locus of the system s obtained, (Fig. 5.33).

losed-loop stability.

a system has an open-loop transfer function A, it was shown

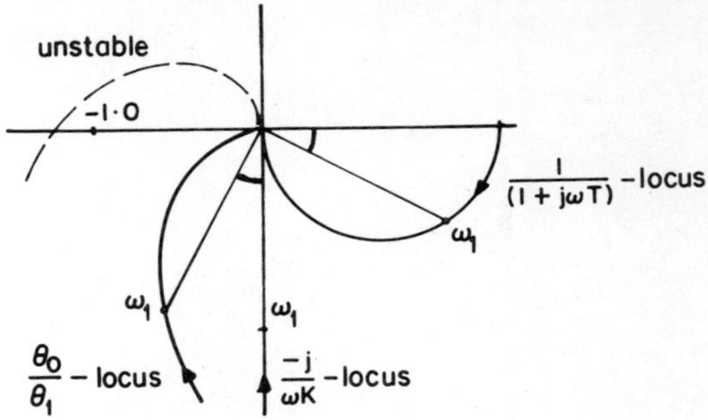

Fig. 5.33.

that the closed-loop transfer function = $1/(1 + 1/A.)$ Hence if th open-loop frequency-response transfer-function plot passes through the point $(-1, j0)$, then $A = -1$. The corresponding closed-loop frequency-response transfer-function, $\theta_0/\theta_1 = \infty$, implying an output for zero input, or instability. Thus a plot of the open-loop transfer function gives an indication of the closed-loop stability of a system. A measure is given by the 'gain margin' and the 'phase margin', as indicated in the frequency-response plot (Fig. 5.34(a)).

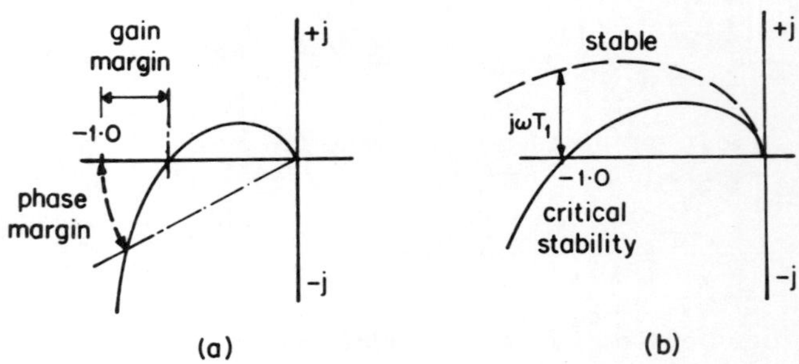

Fig. 5.34.

Improvement of stability is often achieved by the addition of velocity feedback to a system. Thus, in the case considered, th addition of velocity feedback would result in an additional term which would be a function of velocity, say $T_1\ (d\theta/dt)$. The frequency-response definition becomes $\theta(j\omega T_1)$ and this additional term would appear in the denominator of the system transfer function.

The addition may be better appreciated by considering the

inverse of the transfer function, which then becomes:

$$\frac{v_t}{\theta} = j\omega K(1 + j\omega T) + j\omega T_1 .$$

Now the plot of 1/transfer function is termed an Inverse Nyquist diagram and for a stable system this frequency locus will pass *outside* the point (–1, j0).

Thus, to improve the stability of a system, the magnitude of the necessary additional term, $j\omega T_1$, can be estimated directly from an Inverse Nyquist diagram, drawn to scale, Fig. 5.34 (b).

REFERENCES

1. MOULLIN, E.B. (1955). *Electromagnetic Principles of the Dynamo*, Oxford University Press, London.
2. CLAYTON, A.E. *Design of Direct Current Machines*, Pitmans,
3. FITZGERALD, A.E. and KINGSLEY, C. (1961). *Electric Machinery*, 2nd ed. McGraw-Hill, New York.
4. HINDMARSH, J. (1965). *Electrical Machines*, Pergamon Press, Oxford.
5. PUCHSTEIN, A.F. (1961). *The Design of Small Direct Current Motors*, Wiley, New York.
6. THALER, G.J. and BROWN, R.G. *Analysis and Design of Feedback Control Systems*, 2nd ed, McGraw-Hill, New York.
7. WESTWATER, F.L. and WADDELL, W.A. (1951). *An Introduction to Servomechanisms*, The English Universities Press,
8. *General Electric Controlled Rectifier Manual* (1960). G.E.C., Liverpool, New York.
9. *Conference on Power Applications of Controllable Semi-conductor Devices*, I.E.E., Nov. 1965.

EXAMPLES

EXAMPLE 5.1.

State possible reasons why a d.c. shunt generator may fail to excite when it is driven at normal speed on no load.

On full load, a 6 kW, 200V d.c. generator has an efficiency of 76·0%. The armature and shunt-field resistances are 0·5 and 50 Ω respectively. Determine (*i*) the power taken by the machine when run light as a motor from a 200V supply, (*ii*) the shaft output when run as a motor operating on 200V mains and with an armature current of full-load value.

Solution :

First part. The following reasons may be listed, (a) no initial permanent magnetism, (b) the field may be connected the wrong way round, i.e. opposing the permanent magnetism, (c) the field resistance may be greater than the critical value, i.e. the slope of the magnetisation curve, (d) with a new machine it is well to check all circuits, e.g. check that the brushes are in their holders and are bedded down on the commutator.

Second part. As a generator :

$$\text{input} = \frac{6000}{0\cdot76} = 7884\text{ W.}$$

$$I_f = \frac{200}{50} = 4\text{A}, \; I = \frac{6000}{200} = 30\text{A.}$$

Then $I_a = I + I_f = 34\text{A}$ = rated full-load armature current.

Now,
$$I_a^2 R_a = 34^2 \times 0\cdot5 = 578$$
$$I_f^2 R_f = 4^2 \times 50 = 800$$
$$\text{Output} = \underline{6000}$$
$$7378\text{ W.}$$

$$\therefore \quad \text{Rotational loss} = 7884 - 7378 = 506\text{ W.}$$

(*i*) As a motor, running light.

$$\text{Armature input} = \text{rotational loss} + I_a^2 R_a$$
$$200\, I_a = 506 + I_a^2 R_a$$
$$0\cdot5\, I_a^2 - 200\, I_a + 506 = 0$$
$$\therefore \quad I_a = 2\cdot545\text{ A.}$$

Alternatively, since $I_a^2 R_a$ is small, $I_a \approx \dfrac{506}{200} \approx 2\cdot5\text{ A}$

$$\therefore \quad I_a^2 R_a = 2\cdot5^2 \times 0\cdot5 \approx 3\cdot1\text{ W.}$$

$$\text{Corrected } I_a = \frac{509}{200} = 2\cdot545\text{ A.}$$

Hence, power input running light $= (4 + 2\cdot545)\,200$

$= 1309\text{ W.}$

(*ii*) As a motor on full load.

$$I_a = 34\text{A},\ I_f = 4\text{A}, \quad \therefore \quad I_L = 38\text{A}.$$

$$\therefore \quad \text{Input power} = VI_L = 200 \times 38 = 7600\text{W}.$$

$$\text{Output} = 7600 - (34^2 \times 0{\cdot}5 + 800 + 506)$$

$$= 5714\text{W}$$

$$= 7{\cdot}7 \text{ h.p.}$$

EXAMPLE 5.2.

A 220V shunt motor runs at 750 r.p.m. and has $R_a = 0{\cdot}5\,\Omega$ and $I_a = 20$A. (a) Calculate the extra armature resistance to be added to reduce the speed to 500 r.p.m. (b) with the added resistance, the load is now adjusted to give $I_a = 14$A. Calculate the new speed.

Solution :

For a d.c. motor : $V = E + I_a R_a$

(a) $$\therefore \quad E = V - I_a R_a$$

$$= 220 - 20 \times 0{\cdot}5 = 210\text{V}.$$

Now $E \propto N\Phi$ and Φ is constant, (assumed),

$$\therefore \quad \text{New } E = \frac{210}{750} \times 500 = 140\text{V}.$$

$$\text{New } I_a R_a = 220 - 140 = 80\text{V}.$$

$$\text{New } R_a = 4\,\Omega \text{ giving added } R_a = 3{\cdot}5\,\Omega.$$

(b) $$\text{New } I_a = 14\text{A}$$

$$I_a R_a = 14 \times 4 = 56\text{V}$$

$$\text{New } E = 220 - 56 = 164\text{V}$$

$$\text{Then new speed} = \frac{164}{140} \times 500 = 585 \text{ r.p.m.}$$

EXAMPLE 5.3.

A 230V d.c. series motor runs at 600 r.p.m. when taking 110A. If $R_a = 0{\cdot}12\,\Omega$ and $R_f = 0{\cdot}03\,\Omega$, calculate (a) the torque developed, (b) the new speed at reduced load when the machine takes 50A. The magnetization curve gives $2{\cdot}18 \times 10^{-4}$ Wb/A at 110A and $2{\cdot}3 \times 10^{-4}$ Wb/A at 50A.

Solution :

(a)
$$E = V - I_a(R_a + R_{se})$$
$$= 230 - 110 \times 0{\cdot}15 = 213{\cdot}5\text{V}$$
$$2\pi n (T) = EI_a = 213{\cdot}5 \times 110 \text{ where } n = \text{r.p.s.}$$
$$\therefore (T) = \frac{213{\cdot}5 \times 110}{2\pi \times 10} = 375\,\text{N m.}$$

(b)
$$\text{New } E = 230 - 50 \times 0{\cdot}15 = 222{\cdot}5\text{V}$$
$$\Phi_{110} = 2{\cdot}18 \times 10^{-4} \times 110 = 0{\cdot}024\,\text{Wb}$$
$$\Phi_{50} = 2{\cdot}3 \times 10^{-4} \times 50 = 0{\cdot}0155\,\text{Wb}$$

Now $E \propto \Phi N$ or $= k\Phi N \quad \therefore \quad \dfrac{N_1}{N_2} = \dfrac{E_1 \Phi_2}{E_2 \Phi_1}$

$$\therefore \text{ New speed} = \frac{222{\cdot}5 \times 0{\cdot}024 \times 600}{213{\cdot}5 \times 0{\cdot}0155}$$
$$= 969 \text{ r.p.m.}$$

EXAMPLE 5.4.

A d.c. series motor which is rated at 200V, 20kW, has the following magnetic characteristic at 1000 rev/min :

Field current (A)	0	25	50	75	100
Generated e.m.f. (V)	11	144	216	250	278

It has an armature resistance of 0·15 Ω and the series-field resistance is 0·1 Ω.

If the motor is connected to a 200V supply, calculate the speed and shaft torque for currents of 25, 50, 75 and 100% of the full-load value, and plot the corresponding torque-speed curve. The rotational and stray losses may be considered to have a constant value of 1kW and armature reaction may be neglected.

Solution : (see Fig. Ex. 5.4).

Values of motor current, I A =	25	50	75	10
Corresponding resistive drop, (V) $= I(R_a + R_{se}) = I(0{\cdot}15 + 0{\cdot}1) = 0{\cdot}25I$ =	6·25	12·5	18·75	2
On 200V supply, values of induced e.m.f., $E = V - I(R_a + R_{se})$ =	193·75	187·5	181·25	17
SPEED, N in r.p.m. $= E \times 1000/E'$ =	1345	868	726	63
where E' is obtained from magnetic (OCC) characteristic for given I.				

Power input = $200I$ W	=	5000	10000	15000	20000
Copper loss = $I^2(R_a + R_{se}) = 0{\cdot}25I^2$	=	156	620	1410	2500
Total loss = copper loss + 1000W	=	1156	1620	2410	3500
Output (W) = Input (W) − Total loss	=	3844	8380	12590	16500
TORQUE, $(N\text{-}m) = \dfrac{\text{Output (W)} \times 60}{\text{Speed (rpm)} \times 2\pi}$	=	27	92	166	235

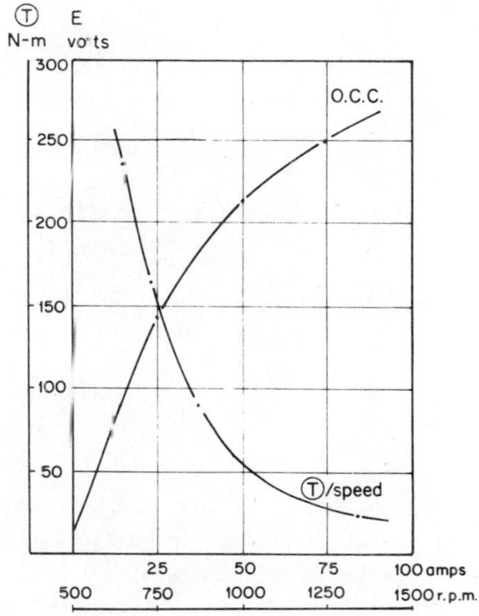

Fig. Ex. 5.4.

EXAMPLE 5.5.

Derive the condition for maximum power output of a d.c. shunt motor, operating on constant supply voltage against a constant gross torque. Assume a linear relationship between field current and flux.

Calculate the speed at maximum output of a 200V d.c. shunt motor under the above conditions. The machine has armature resistance of 0·5 Ω and takes an armature current of 20A when the speed is 570 r.p.m.

Solution: At constant torque, max. o/p occurs at max. speed.

$$\text{Power output} = \omega Ⓣ = VI_a - I_a^2 R_a.$$

$$\text{Let } Ⓣ = 1/K = \text{constant}$$

$$\therefore \quad d\omega/dI_a = K(V - 2I_a R_a) = 0 \text{ for maximum.}$$

$$\therefore \quad V = 2I_a R_a \text{ or } I_a = \frac{V}{2R_a} \text{ is the condition.}$$

Numerical part:

$$\text{At max. o/p: } I_a = \frac{V}{2R_a} = \frac{200}{2 \times 0{\cdot}5} = 200\text{A.}$$

$$\therefore \quad E_b = V - I_a R_a = 200 - 200 \times 0{\cdot}5$$
$$= 100\text{V.}$$

$$\text{At 570 r.p.m.} \quad E_b = 200 - 20 \times 0{\cdot}5 = 190\text{V.}$$

$$\text{Now, } Ⓣ \propto \Phi I_a \propto I_f I_a = \text{constant.} \quad \therefore \frac{I_{a_1}}{I_{a_2}} = \frac{I_{f_2}}{I_{f_1}}.$$

$$\text{Also, } E_b = k\Phi N \propto I_f N \quad \therefore \quad \frac{E_{b_1}}{E_{b_2}} = \frac{I_{f_1} N_1}{I_{f_2} N_2} = \frac{N_1 I_{a_2}}{N_2 I_{a_1}}$$

$$\therefore \quad N_1 = \frac{E_{b_1}}{E_{b_2}} \cdot \frac{I_{a_1}}{I_{a_2}} \cdot N_2 = \frac{100}{190} \times \frac{200}{20} \times 570 = \underline{3000 \text{ r.p.m.}}$$

EXAMPLE 5.6.

The armatures of two 200V, 15kW separately-excited d.c. generators, *A* and *B*, are connected in parallel to supply a common load. The open-circuit e.m.f's of the two generators are 210V and 200V respectively, and their armature resistances are 0·12Ω and 0·1Ω respectively. Their internal characteristics may be considered to be linear, and the percentage fall in generated e.m.f. on full load due to armature reaction and change in speed is 6% for *A* and 5% for *B*. Calculate the common armature voltage and the output of each generator, when they share a load of 120A.

Solution:

For each machine, full-load current = $15 \times 10^3/200 = 75$A.
Terminal voltage = O/C voltage – (volt drop due to armature reaction and speed change) – $I_a R_a$.

$\therefore$ For generator A:

$$V_A = 210 - \left(\frac{6}{100} \times 210 \times \frac{I_A}{75}\right) - 0{\cdot}12 I_A$$
$$= 210 - 0{\cdot}288 I_A.$$

Similarly: $V_B = 200 - \left(\frac{5}{100} \times 200 \times \frac{I_B}{75}\right) - 0{\cdot}1\,I_B$

$$= 200 - 0{\cdot}233\,I_B.$$

Under operating conditions: $V_A = V_B$ and $I_A + I_B = 120$.
Substituting for V_A and V_B,

$$210 - 0{\cdot}288\,I_A = 200 - 0{\cdot}233\,I_B.$$

$$\therefore \quad 0{\cdot}288\,I_A - 0{\cdot}233\,I_B = 10.$$

Also $0{\cdot}233\,I_A + 0{\cdot}233\,I_B = 27{\cdot}9$, (by multiplying across the current relationship by 0·233).

Solving for I_A, we have: $0{\cdot}521\,I_A = 37{\cdot}9$

$$\therefore \quad I_A = \underline{72{\cdot}9\text{A}}$$

$$I_B = 120 - 72{\cdot}9 = \underline{47{\cdot}1\text{A}}.$$

The common terminal voltage is given:

$$V_A = V_B = 200 - 0{\cdot}233\,I_B = \underline{189\text{V}}.$$

EXAMPLE 5.7.

A 440V d.c. shunt motor runs at 850 r.p.m. at no load, at which it takes an armature current of 1·2A. The effective armature resistance is 0·8 Ω and the field resistance is 400 Ω. The motor has the following magnetic characteristic when driven at a constant speed of 850 r.p.m. (Fig. Ex. 5.7).

Field current (A)	0·25	0·5	0·75	1·0	1·25
Generated e.m.f. (V)	193	333	398	431	451

Neglecting variations of rotational losses with speed, calculate the resistance which must be added to the field circuit in order to obtain an output of 10 h.p. at a speed of 1000 r.p.m.
Solution:

Required output $= 10 \times 746 = 7460\text{W}$.

No-load rotational loss $= VI_0 - I_0^2 R_a$

$$= 440 \times 1{\cdot}2 - (1{\cdot}2)^2 \times 0{\cdot}8$$

$$= 527\text{W}.$$

Armature copper loss $= I_a^2 R_a = 0{\cdot}8\,I_a^2$.

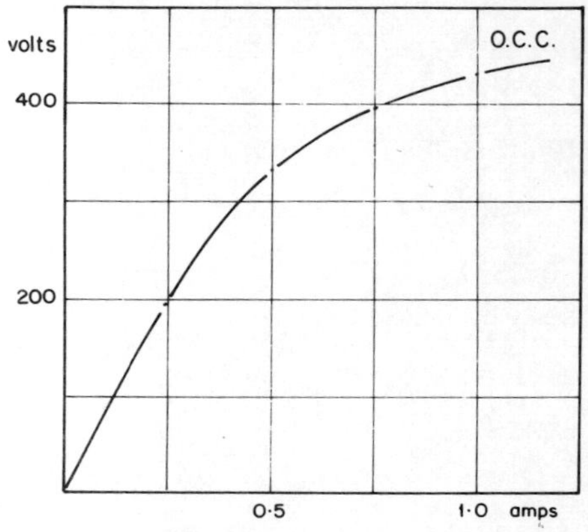

Fig. Ex. 5.7.

$$\text{Now armature current, } I_a = \frac{\text{output + no load loss + arm. cu. loss}}{\text{arm. applied volts}}$$

$$\therefore \quad I_a = \frac{7460 + 527 + 0{\cdot}8\, I_a^2}{440}$$

$$0{\cdot}8\, I_a^2 - 440\, I_a + 7987 = 0$$

$$\therefore \quad I_a = 18{\cdot}8\text{A}.$$

The back e.m.f. at 10 h.p. output $= 440 - 18{\cdot}8 \times 0{\cdot}8$
$= 425\text{V}.$

This is at 1000 r.p.m. At 850 r.p.m.

$$E_b = 425 \times \frac{850}{1000} = 361\text{V}.$$

From the O.C.C. graph, $I_f = 0{\cdot}581\text{A}.$

$$\therefore \quad R_f = \frac{440}{0{\cdot}581} = 757\,\Omega.$$

Hence the increase in R_f required $= 757 - 400 = \underline{375\,\Omega}$.

EXAMPLE 5.8.

A four-pole d.c. machine has a wave-wound armature with 888 conductors. The brushes are displaced 5° mech. from the magnetic neutral axis. Calculate (a) the demagnetising and the cross-magnetising ampere-turns/pole if the armature terminal current is 90A, (b) the turns/pole of a compensating winding, given pole-arc/pole-pitch = 0·7.

Solution:

$$\text{Total ampere-turns/pole} = \frac{1}{2}\left(\frac{ZI}{2pa}\right).$$

$$\text{(a)} \qquad \theta^\circ \text{ electrical} = p(\text{degrees mech.})$$

$$= 2 \times 5 = 10^\circ \text{ elect.}$$

$$\therefore \text{ Demagnetizing AT/pole} = \frac{4\theta}{360}\left(\frac{1}{2}\frac{ZI}{2pa}\right) = \frac{40}{360}\left(\frac{888 \times 90}{2 \times 8}\right)$$

$$= 555 \text{ AT/pole.}$$

$$\text{Cross-magnetizing AT/pole} = \left(1 - \frac{40}{360}\right)\left(\frac{888 \times 90}{2 \times 8}\right) = 4440 \text{ AT/pole.}$$

$$\text{(b) Compensating AT/pole} = \frac{\text{pole-arc}}{\text{pole-pitch}}\left(\frac{1}{2}\frac{ZI}{2pa}\right)$$

$$= 0{\cdot}7\ \frac{888 \times 90}{2 \times 8}.$$

Since the winding is connected in series with the armature terminal, the current in the winding = 90A.

$$\therefore \text{ Compensating turns/pole} = 0{\cdot}7\left(\frac{888}{2 \times 8}\right)$$

$$= 39 \text{ turns/pole.}$$

EXAMPLE 5.9.

The field winding of a four-pole d.c. machine has 500 turns per pole and each turn carries a current of 3A. The pole arc is 70% of the pole pitch. The armature of the machine is wave-wound, with 43 slots and 12 conductors per slot. Calculate approximately the air-gap magneto-motive force under the centre and both tips of a pole when the armature current is 30A.

Describe the principal assumptions made in performing the calculations and discuss their effect on the validity of the results.

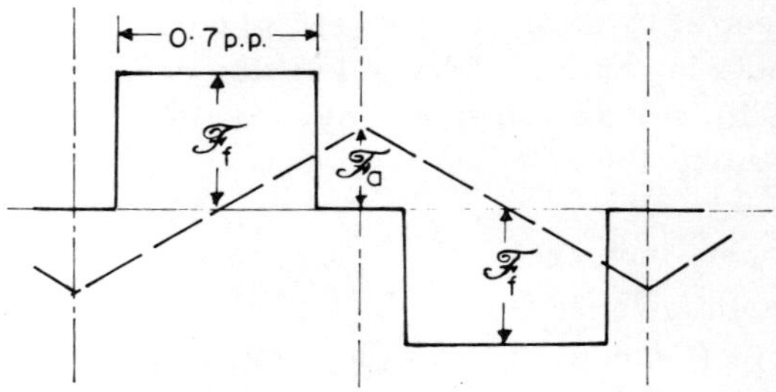

Fig. Ex. 5.9.

Solution :

(*i*) Field coil m.m.f., $\mathcal{F}_f = 3 \times 500 = 1500$ AT.

This is the m.m.f. under the pole centre, (Fig. Ex. 5.9).

Now $$\mathcal{F}_a = \frac{1}{2} \cdot \frac{I_a Z_a}{2p}$$

where I_a is the arm. conductor current, p is the pairs of poles, and Z_a is the total number of armature conductors. Then we have :

$$I_a = I/2 = 15\text{A, (wave winding).}$$
$$Z_a = 43 \times 6 \times 2 = 516$$
$$p = 2$$
$$\therefore \mathcal{F}_a = \frac{1}{2} \times \frac{15 \times 516}{4} = 967 \text{ AT.}$$

Under the pole tips, (Fig. Ex. 5.9), the m.m.f's are:

$$\mathcal{F}_f + 0{\cdot}7\,\mathcal{F}_a = 1500 + 676 = 2176 \text{ AT.}$$
$$\mathcal{F}_f - 0{\cdot}7\,\mathcal{F}_a = 1500 - 676 = 824 \text{ AT.}$$

(*ii*) The principal assumptions made are:

1. All the m.m.f. is expended in the airgap,
2. Fringing fluxes are neglected,
3. Effect of teeth on the gap reluctance neglected.

Further discussion on p.118.

EXAMPLE 5.10.

A 1000 r.p.m. d.c. separately-excited generator has the following constants; $R_a = 0{\cdot}3\,\Omega$, $L_a = 0{\cdot}08\,\text{H}$, $R_f = 100\,\Omega$, $L_f = 20\,\text{H}$, slope of magnetization curve is 250 V/A at 500 r.p.m. (a) Calculate the transfer function on open circuit when running at normal speed, (b) if a load of $R = 9{\cdot}5\,\Omega$ and $L = 0{\cdot}9\,\text{H}$ is connected to the armature, calculate the instantaneous armature current when 50V d.c. is suddenly applied to the field. Assume a 2% drop in rated speed due to the load and a linear e.m.f./speed relation.

Solution:

(a) Transfer function on o/c $= \dfrac{\text{armature volts}}{\text{field volts}}$.

In standard notation, $v_f = i_f\,(R_f + L_f\,p)$

$$\therefore \quad i_f = \frac{v_f}{R_f}\left(\frac{1}{1 + T_f\,p}\right)$$

$$e_0 = K_g i_f = \frac{K_g}{R_f}\left(\frac{1}{1 + T_f\,p}\right) v_f\,.$$

Now, $T_f = L_f/R_f = 20/100 = 0{\cdot}2$ sec.

At rated speed, $K_g/R_f = \left(\dfrac{250 \times 1000}{100 \times 500}\right) = 5$

$$\therefore \quad \text{Transfer function} = e_0/v_f = 5\left(\frac{1}{1 + 0{\cdot}2p}\right).$$

(b) With load, 2% drop in speed and $V_f = 50\text{V}$, gives:

$$e_a = \frac{5 \times 98 \times 50}{100}\left(\frac{1}{1 + 0{\cdot}2p}\right)$$

$$i_a = \frac{e_a}{R_a + R}\left(\frac{1}{1 + T_a\,p}\right) \text{ where } T_a = \frac{L_a + L}{R_a + R} = 0{\cdot}1 \text{ sec.}$$

$$\therefore \quad i_a = \frac{250 \times 98}{100 \times 9{\cdot}8}\left(\frac{1}{1 + 0{\cdot}2p}\right)\left(\frac{1}{1 + 0{\cdot}1p}\right)$$

$$= 25\left(\frac{1}{1 + 0{\cdot}2p} \times \frac{1}{1 + 0{\cdot}1p}\right)$$

$$\therefore \quad i_a = I + A\epsilon^{-t/T_f} + B\epsilon^{-t/T_a}.$$

Now *either*

At $t = 0,\ i = 0 \quad \therefore \quad 0 = 25 + A + B$

$di/dt = 0 \quad \therefore \quad 0 = -5A - 10B$

$A = -50$

$B = 25$

$$\therefore \quad i_a = 25\,(1 - 2\epsilon^{-5t} + \epsilon^{-10t})\,\text{A}.$$

Or using partial fractions :

$$i_a = I\left(1 - \frac{T_f}{T_f - T_a}\,\epsilon^{-t/T_f} - \frac{T_a}{T_a - T_f}\,\epsilon^{-t/T_a}\right)$$

$$= 25\left(1 - \frac{0{\cdot}2}{0{\cdot}2 - 0{\cdot}1}\,\epsilon^{-t/0\cdot2} - \frac{0{\cdot}1}{0{\cdot}1 - 0{\cdot}2}\,\epsilon^{-t/0\cdot1}\right)$$

$$= 25\,(1 - 2\epsilon^{-5t} + \epsilon^{-10t})\,\text{A}.$$

EXAMPLE 5.11.

The field voltage of a separately-excited d.c. motor is supplied by an amplifier of gain A and output resistance R. The armature is supplied from a constant-current source. The combined inertia of armature and load is J and the frictional damping is D. Derive expressions for the shaft torque and output speed, in terms of the system time-constants and the input voltage to the amplifier.

If the output speed is monitored by a tachogenerator to provide proportional negative-feedback voltage to the input, derive a modified solution for the output speed for this case. Comment on the factors which control the performance of the system.

Solution :

It is generally wise to draw a diagram of the situation described before commencing the solution, Fig. Ex. 5.11.

Now Ⓣ $= i_a i_f M_d = K_a i_f$ where $K_a = M_d i_a$, (i_a is constant)

$$i_f = \frac{v_f}{R + R_f}\left(\frac{1}{1 + T_f p}\right) \text{ where } T_f = L_f/(R + R_f)$$

$$\therefore \quad \textcircled{T} = \frac{K_a}{R + R_f}\left(\frac{1}{1 + T_f p}\right) v_f \text{ and } v_f = Ae$$

where e = input voltage to amplifier.

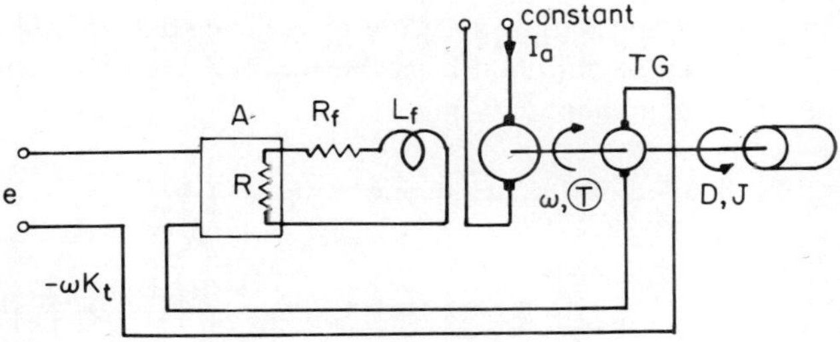

Fig. Ex. 5.11.

Considering speed, in mech. radians/sec:

$$J \cdot d\omega/dt + D\omega = \textcircled{T}$$

more briefly,

$$\omega(Jp + D) = \textcircled{T}$$

$$\therefore \omega = \textcircled{T}\frac{1}{Jp + D} = \frac{K_a}{D(R + R_f)}\left[\frac{1}{(1 + T_f p)(1 + T_m p)}\right] Ae$$

where $T_m = J/D$.

Introducing negative feedback $= -K_t \omega$

$$\therefore \omega = \left[\frac{K_1}{(1 + T_f p)(1 + T_m p)}\right](e - K_t \omega) \text{ where } K_1 = \frac{AK_a}{D(R + R_f)}.$$

Collecting terms in ω:

$$\omega\left[1 + \frac{K_1 K_t}{(1 + T_f p)(1 + T_m p)}\right] = \left[\frac{K_1}{(1 + T_f p)(1 + T_m p)}\right] e$$

$$\therefore \omega = \left[\frac{K_1}{(1 + T_f p)(1 + T_m p) + K_1 K_t}\right] e$$

$$= \left[\frac{K_1}{p^2(T_f T_m) + p(T_f + T_m) + (1 + K_1 K_t)}\right] e$$

$$= \frac{K_1}{(1 + K)}\left[\frac{1}{p^2\left(\dfrac{T_f\, T_m}{1 + K}\right) + p\left(\dfrac{T_f + T_m}{1 + K}\right) + 1}\right] e$$

where $K = K_1 K_t = K_t\left\{\dfrac{AK_a}{D(R + R_f)}\right\}$.

The equation for ω represents a linear second-order differential equation with a solution as a damped sinusoid. The natural frequency is obtained from the roots of the quadratic equation in the denominator, thus:

$$\text{Undamped natural frequency, } \omega_n = \left(\frac{1 + K}{T_f\, T_m}\right)^{\frac{1}{2}}$$

$$\text{Damping factor, } a = \frac{1}{2}\left(\frac{1}{T_f} + \frac{1}{T_m}\right)$$

and of course, $\omega_d = (\omega_n^2 - a^2)^{\frac{1}{2}}$.
The relative damping or Damping Ratio, $\zeta = \dfrac{a}{\omega_n}$.

Note that the value of T_f depends on the output resistance of the amplifier. Also, increasing A increases ω_n but decreases ξ. A value of ξ between 0·5 and 0·7 is generally satisfactory.

6 The synchronous machine

All large quantities of electrical energy are generated by synchronous machines. The synchronous machine may operate as a generator or a motor, depending on the nature of the input energy.

The armature winding, usually in three phases, is located on the stator and the field winding on the rotor. This arrangement has advantages, viz:

(a) The armature terminals, (three lines and neutral), are stationary. Insulation problems and problems of heavy current connections are reduced. For a given depth of slot, more winding space is available for copper and insulation on the stator side of the airgap.

(b) Only two slip-rings are required to supply the relatively low voltage, low power d.c. supply to the field coils. Some recent designs dispense with slip-rings entirely by locating the exciter armature with solid-state rectifiers on the rotor, to provide the d.c. excitation current. Fig. 6.1.

The machine is classified according to the type of rotor, namely 'salient-pole' or 'non-salient-pole'. Non-salient-pole, or cylindrical rotors are used in high speed machines, commonly styled 'turbo-alternators'. At the high speed, (3000 or 3600 r.p.m.), only two poles are required. The field coils are distributed in slots on the mechanically strong rotor forging, providing a robust construction necessary at high speeds to combat the considerable centrifugal forces, (Fig. 6.2(a)). The rotor is of small diameter and long axial length. Peripheral speeds up to

140 m/s are quoted.

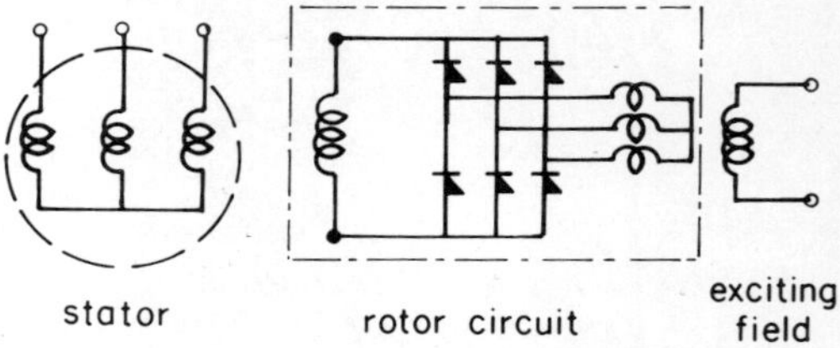

Fig. 6.1.

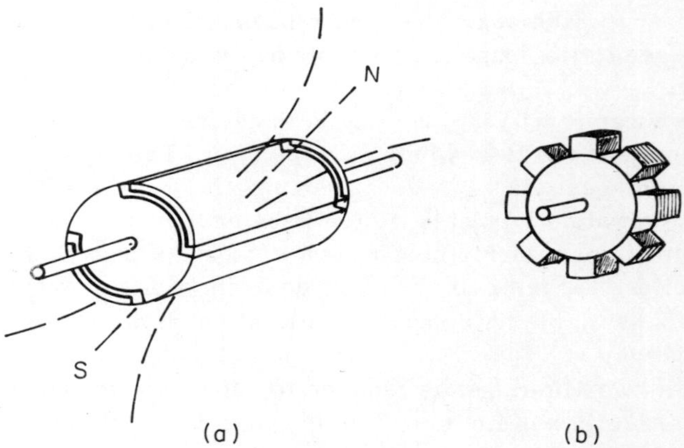

Fig. 6.2.

The principal design problems concern:

(*i*) Cooling. Various methods of forced cooling are used. Water cooling for the stator, where hollow conductors are possible and also CO_2 or hydrogen gas, (95% pure) are employed.

(*ii*) The effect of the end-bells or strengthening bands which are added to the rotor.

(*iii*) The forced circulation of cooling gas involves strategic location of cooling ducts. These can introduce undesirable noise.

The salient-pole rotor is used on slow speed machines and generally where four or more poles are necessary. Fig. 6.2(b). The rotor design is thus of large diameter, small axial length and the machines have been styled 'flywheel' type. Peripheral speeds may reach 60–70 m/s. For hydro-electric machines, sudden disconnection of load can cause dangerous 'run-away' speeds and the design should allow for this.

Interesting features of design in hydro-electric sets are variable-pitch water turbines and the employment of the generator to run as a synchronous motor to pump water back to high levels during periods of light loading of the associated supply system.

Synchronous motors have a wide range of application, ranging from large turbo-electric marine drives, power-factor correction in sub-stations, to small exact-speed drives. But they are inherently not self-starting (p. 64).

The rotating field principle

Consider two polarised fluxes, each of constant magnitude Φ, superimposed on a common axis. If they are rotated at constant speed, ω, in opposite directions, the component of each on the original axis $= \Phi \cos \omega t$ and the resultant flux $= 2\Phi \cos \omega t$ along the axis, Fig. 6.3. Hence the combined effect is a sinusoidal alternating flux along the common axis, of frequency $\omega/2\pi$.

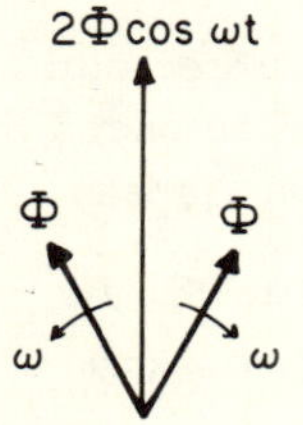

Fig. 6.3.

Conversely, an axial alternating flux, $\Phi_m \cos \omega t$ can be represented by two component conjugate fluxes, each of constant magnitude $\Phi_m/2$ and rotating in opposite directions at ω radians/sec, i.e. synchronous with the flux alternations.

The two-phase system

Consider two coils, physically at 90° and supplied from a balanced two-phase supply, Fig. 6.4. Then, instantaneous coil fluxes can be written :

$$\phi_1 = \Phi_m \cos \omega t$$
$$\phi_2 = \Phi_m \cos(\omega t - 90°) = \Phi_m \sin \omega t.$$

Resolving physically, resultant flux $\phi_r = \sqrt{(\phi_1^2 + \phi_2^2)}$

$$= \Phi_m$$

Also $\theta = \omega t.$

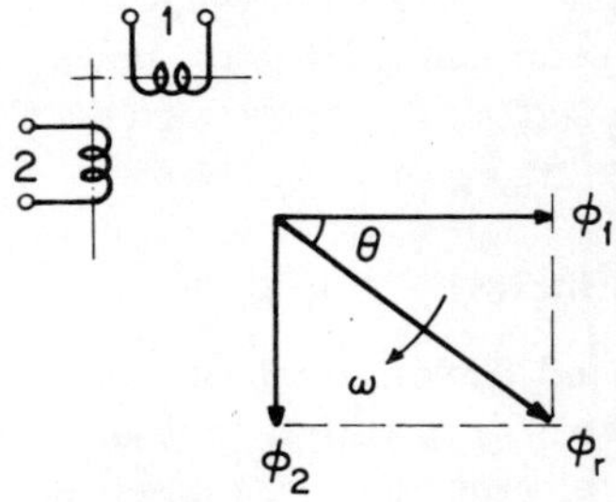

Fig. 6.4.

Hence the two-phase system produces a synchronous field, of constant magnitude = maximum flux/phase.

The three-phase case

Using a notation as above, three coils located on axes physically at 120° and supplied from a balanced three-phase supply, will produce instantaneous axial fluxes:

$$\phi_1 = \Phi_m \cos \omega t$$
$$\phi_2 = \Phi_m \cos(\omega t - 120°)$$
$$\phi_3 = \Phi_m \cos(\omega t - 240°).$$

The resultant flux produced by this system is determined by resolving the components with respect to physical axes, Fig. 6.5. Hence we have:

Sum of horizontal components:

$$\phi_h = \Phi_m[\cos \omega t - \cos(\omega t - 120°) \cos 60° - \cos(\omega t - 240°) \cos 60°]$$

$$= \frac{3}{2} \Phi_m \cos \omega t.$$

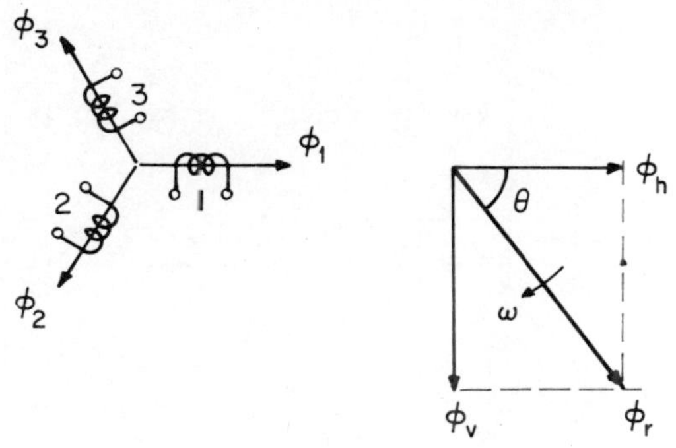

Fig. 6.5.

Sum of vertical components :

$$\phi_v = \Phi_m[0 - \cos(\omega t - 120°)\sin 60° + \cos(\omega t - 240°)\sin 60°]$$

$$= \frac{-3}{2}\Phi_m \sin \omega t.$$

$\therefore$ Resultant flux: $\phi_r = \sqrt{(\phi_v^2 + \phi_h^2)} = \frac{3}{2}\Phi_m$

and $\theta = \omega t.$

Thus the resultant flux is synchronous and of constant magnitude = 3/2 times the maximum flux/phase.

These principles are important to the understanding of synchronous motor operation. A d.c. polarised rotor will tend to align with the stator synchronous field and for steady-state operation will maintain a fixed misalignment, δ, termed the load angle, consistent with the load torque. The principles are important also in relation to induction motor operation, considered later.

Note that the cases considered assume concentrated coils.

The three-phase system with distributed winding

The m.m.f. pattern produced by a distributed three-phase winding with balanced three-phase currents is shown in Fig. 6.6. Instantaneous phase-m.m.f.'s are combined to form this resultant m.m.f. distribution. The resultant m.m.f. varies between the two limiting trapezoidal shapes shown. Note that the axis of the resultant m.m.f. form moves 30° in space for a time interval of 30° electrical. In other words, the m.m.f. wave moves synchronously with the time vectors.

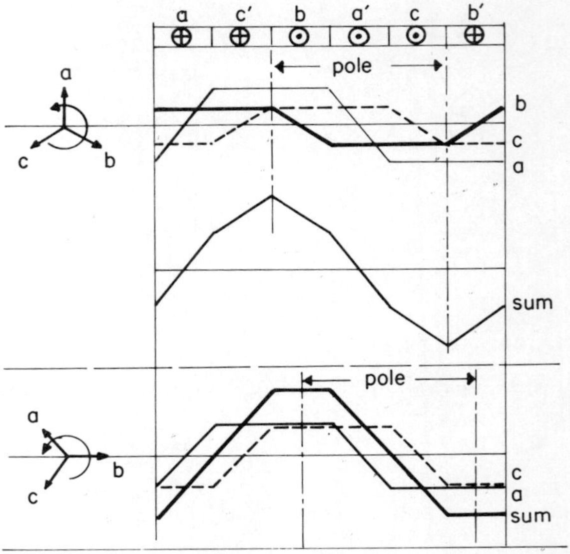

instantaneous m.m.f. distribution patterns

Fig. 6.6.

For a phase current I(r.m.s.), the peak m.m.f./phase

$$\mathcal{F} = \frac{\pm\sqrt{2} \cdot I \cdot Z_p/2}{2p} .$$

Hence the average of the three-phase trapezoidal forms

$$= \frac{(2\mathcal{F} + \sqrt{3} \cdot \mathcal{F})}{2} = 1{\cdot}866\,\mathcal{F}.$$

Note the value 3/2 (phase max.) obtained for concentrated coils, which shows that distribution of the winding has the effect of raising the maximum flux density.

Neglecting the higher harmonics, the trapezoidal waveforms are approximated to the fundamental sine-wave. By Fourier analysis the amplitude of the phase fundamental

$$= \frac{4}{\pi}\mathcal{F} \text{ and for the three phases } = \frac{6}{\pi}\mathcal{F}.$$

∴ Amplitude of the fundamental for the resultant m.m.f. waveform:

$$= \frac{6}{\pi}\mathfrak{F} = \frac{6}{\pi} \cdot \frac{\sqrt{2} \cdot I \cdot Z_p/2}{2p}$$

$$= 1{\cdot}35 \frac{I Z_p}{2p}.$$

Steady-state performance

Performance characteristics of a.c. machines are commonly related to the concept of an equivalent circuit of the machine. This can take the form of an ideal generator and an associated equivalent impedance. Alternatively, inductance coefficients may be defined as in the Generalised Theory considered later. In both cases complex electro-magnetic conditions are interpreted in terms of circuit components.

Synchronous impedance

In this discussion balanced conditions are assumed and generally, 'per phase' notation is implied.

When a load is connected to the terminals of an a.c. generator, the terminal voltage on open-circuit E changes to the on-load terminal voltage V. The condition can be expressed simply as:

$$V = E - IZ_s$$

where Z_s is the 'synchronous impedance' of the machine. The nature of Z_s is developed below.

The machine has a rated terminal voltage V at full load and the 'voltage regulation' is defined as the 'rise' in terminal voltage when full load at rated p.f. is removed from the machine, viz., as a percentage:

$$\text{Regulation} = \frac{E - V}{V} \cdot 100\%.$$

The change in terminal voltage depends upon:

(*i*) The ohmic resistance of the armature winding. Commonly at power frequencies, the d.c. resistance is used, but the a.c. value is more correct, allowing for skin-effect. It may be determined from the watts calculated from a s.c. test on the machine.

(*ii*) Leakage reactance, X_1. This accounts for a loss of voltage of the nature of a reactance drop and is due to leakage fluxes at slot tips and at winding end-connections, (cf. transformer leakage reactance).

(*iii*) Armature reaction. The influence of armature m.m.f. upon the airgap flux affects the generated voltage. In a.c. machines, the effect is governed by the armature current and the power factor of the load. The essential effects can be deduced from Fig. 6.7.

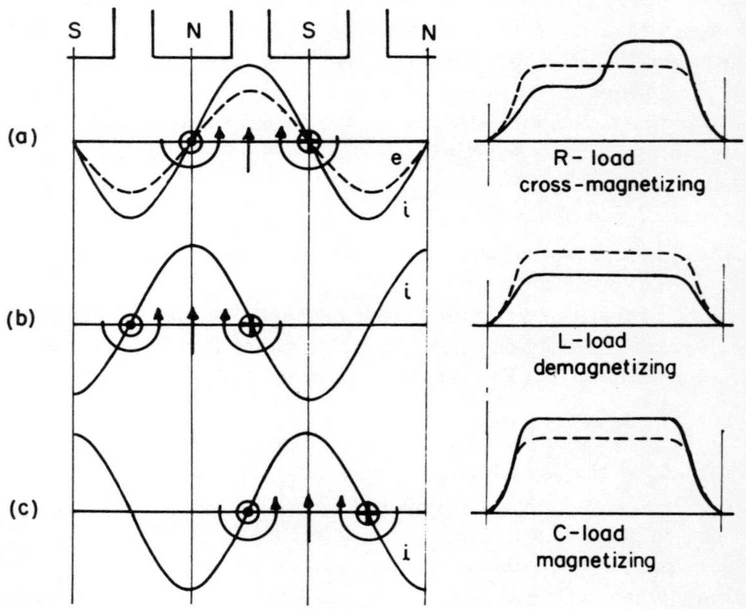

Fig. 6.7.

With the relative pole-positions shown, the armature m.m.f. waveforms correspond to in-phase armature current, i.e. a pure resistance load on a generator. (a). The effect of the m.m.f. on the pole is cross-magnetizing, causing demagnetization of the leading pole-tip and direct magnetization of the trailing pole-tip. The distortion of the flux leads to the development of torque, and illustrates that a resistance load demands power from the machine

With a pure inductance load, current lags voltage by 90° elect. and the position of maximum armature m.m.f. occurs 90° elect.

later, acting directly on the pole axis. (b). The armature m.m.f. effect in this case is thus direct demagnetizing. Similarly, with a pure capacitance load, (c) current leads voltage by 90° elect. and the armature m.m.f. peak occurs 90° elect. previous to that shown at (a). Then the result is a direct magnetizing action on the pole.

Thus with the assumption of sinusoidal m.m.f. distribution, component space-oriented m.m.f's are combined in the manner of time-variable phasors, at the appropriate armature current phase-angle. Further, assuming magnetic linearity and uniform airgap, a sinusoidal time-varying flux is produced by the synchronous space-oriented effective field m.m.f. Hence time and space oriented diagrams may be related in one diagram.

The phasor diagram for the synchronous generator can now be drawn as follows: Fig. 6.8.

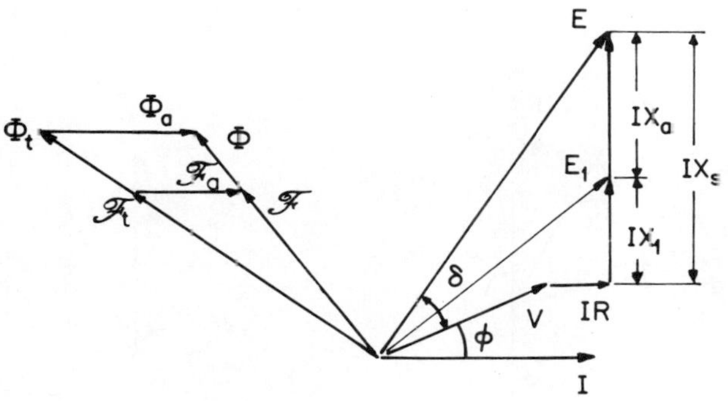

Fig. 6.8.

Working per phase, load conditions are defined as V, I and ϕ. Allowing for the IR and IX_1 voltage drops, an internal voltage E_1 is determined. This is termed the 'airgap voltage'. It is assumed to be generated by the resultant gap-flux Φ. But Φ is produced by a proportional effective field m.m.f., $\mathcal{F}$. Combining $\mathcal{F}$ with the armature m.m.f. $\mathcal{F}_a$, with due regard to phase, open-circuit total field m.m.f. $\mathcal{F}_t$ is determined. This causes flux Φ_t, which induces open-circuit voltage E.

An allowance Φ_{a_1} may be made for armature leakage flux. The significance of misalignment, or 'load angle', as a shift in the effective flux axis, is indicated. The load angle δ, in electrical degrees, occurs between the E and V phasors. The load angle can be observed and measured by stroboscopic means at the machine shaft in mechanical degrees, ψ (where $\delta = p\psi$).

With the assumptions of magnetic linearity and of uniform airgap, it is seen that the effect of armature reaction is equivalen to an additional reactance X_a. Hence X_a is termed 'the reactanc equivalent of armature reaction'. Also, the effective reactance of the machine is $(X_1 + X_a) = X_s$, termed the synchronous reactance.

Thus synchronous impedance $Z_s = \sqrt{(R^2 + X_s^2)}$.

Operating as a synchronous motor, V' is the impressed termina voltage and armature current is supplied to the machine. Armatur reaction effects are reversed in polarity and the phasor diagram for motor operation is shown in Fig. 6.9(a). Note that, to retain the bus-bar voltage as reference for transition from generator to motor operation, the phasor diagram is often shown as in Fig. 6.9(b).

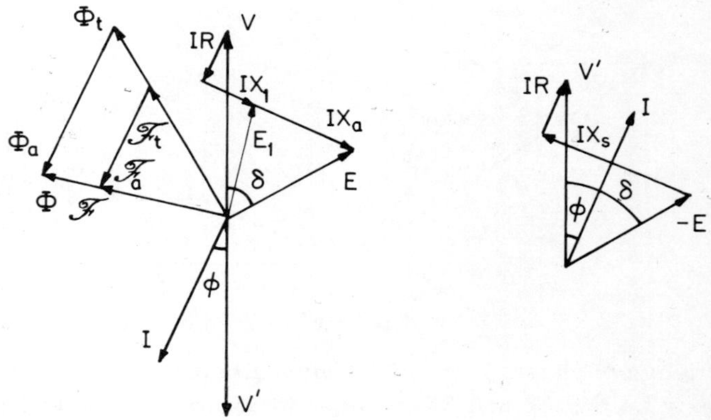

(a) (b)

Fig. 6.9.

The simple equivalent circuit, Fig. 6.10, leads to valuable assessment of machine performance, including the salient-pole case when a greater measure of approximation is involved.

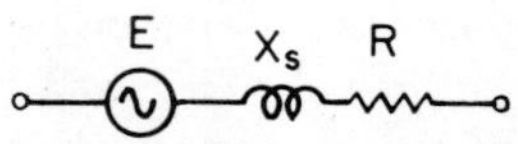

Fig. 6.10.

The power/angle relation

Important conclusions regarding the steady-state behaviour of the synchronous machine on load are derived as follows:
For generator operation, working per phase:

$$\text{Power output, } P_0 = VI\cos\phi.$$

Referring to the phasor diagram, drawn with V as reference phasor Fig. 6.11.

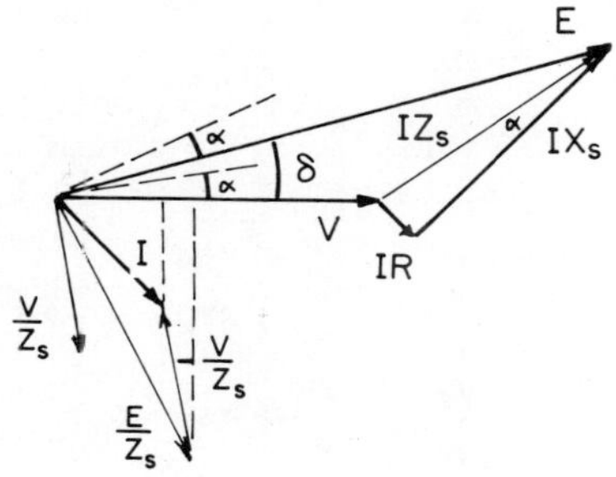

Fig. 6.11.

$$\mathbf{I} = \frac{\mathbf{E} - \mathbf{V}}{\mathbf{Z}_s} = \frac{E\angle\delta - V\angle 0}{Z_s\angle 90 - \alpha} = \frac{E}{Z_s}\angle\delta + \alpha - 90 - \frac{V}{Z_s}\angle\alpha - 90$$

where $\alpha = \tan^{-1} R/X_s$.

Now the in-phase components of the current combine to give:

$$I\cos\phi = \frac{E}{Z_s}\sin(\delta + \alpha) - \frac{V}{Z_s}\sin\alpha$$

$$\therefore \quad P_0 = VI\cos\phi = \frac{VE}{Z_s}\sin(\delta + \alpha) - \frac{V^2}{Z_s}\sin\alpha.$$

A similar expression is obtained for power input, viz:

$$\begin{aligned} P_i &= \text{output} + \text{losses} \\ &= VI\cos\phi + I^2R \\ &= EI\cos(\phi + \delta). \end{aligned}$$

Taking E as the reference phasor:

$$\mathbf{I} = \frac{E\angle 0 - V\angle -\delta}{Z_s\angle 90 - \alpha} = \frac{E}{Z_s}\angle\alpha - 90 - \frac{V}{Z_s}\angle -\delta + \alpha - 90.$$

Then $$I \cos(\phi + \delta) = \frac{V}{Z_s} \sin(\delta - \alpha) + \frac{E}{Z_s} \sin \alpha$$

and $$P_i = \frac{VE}{Z_s} \sin(\delta - \alpha) + \frac{E^2}{Z_s} \sin \alpha.$$

For synchronous motor operation

Applying like principles to the phasor diagram of the motor it is shown:

$$P_o = \frac{VE}{Z_s} \sin(\delta + \alpha) - \frac{E^2}{Z_s} \sin \alpha$$

$$P_i = \frac{VE}{Z_s} \sin(\delta - \alpha) + \frac{V^2}{Z_s} \sin \alpha.$$

With large modern machines, $R \ll X_s$ and R may be considered negligible. In this case the several expressions for power reduce to:

$$P = \frac{VE}{X_s} \sin \delta \text{ watts/phase.}$$

(a) Note that with the conditions of constant rated V and constant E, δ is the only variable in the power equations, and the important characteristic termed the power/angle relation is obtained, Fig. 6.12.

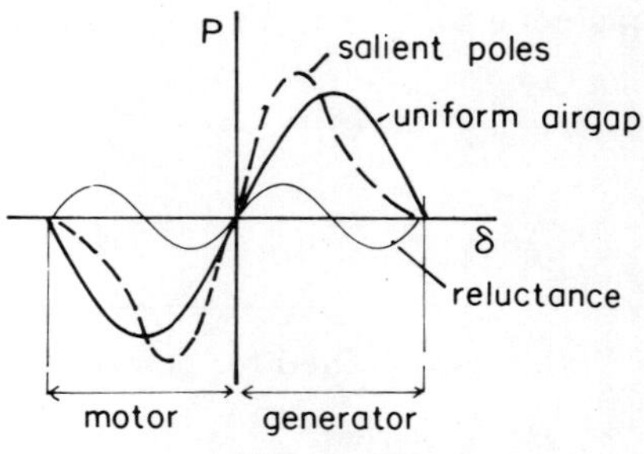

Fig. 6.12.

Also $$P_{max} = \frac{VE}{X_s}$$

and this occurs at $\delta = 90°$.

If R is not neglected, then maximum electrical power occurs, for the generator when $\sin(\delta + \alpha) = 1$ or $\delta = (90 - \alpha)$ and for the motor when $\sin(\delta - \alpha) = 1$ or $\delta = (90 + \alpha)$.

(b) Consider the mode of operation with constant power, variable excitation, and normal constant bus-bar voltage. For generator operation, Fig. 6.13.

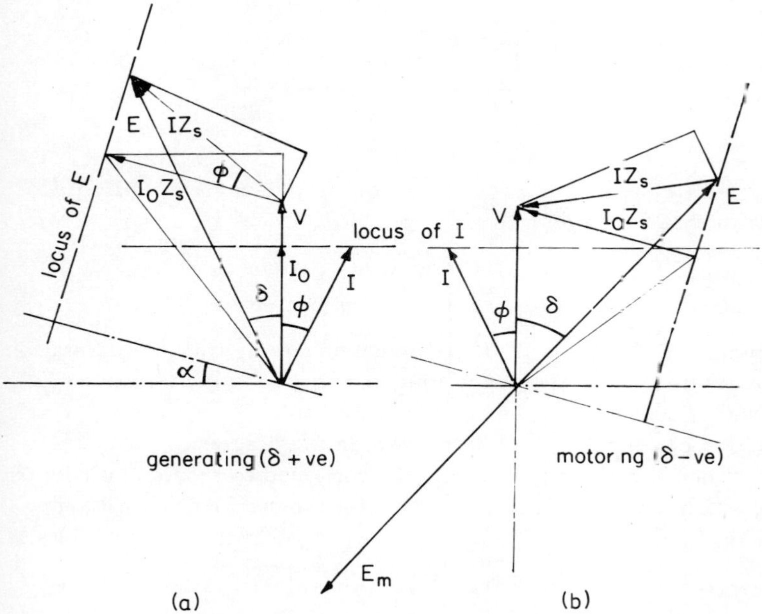

Fig. 6.13.

$$\text{Power} = VI\cos\phi = \text{constant.}$$

$$\text{Since } V = \text{constant} \therefore \quad I\cos\phi = \text{constant}$$

$$= \text{unity p.f. current, } I_0 .$$

Hence the locus of I is perpendicular to the V phasor. The experimental plot of I vs. I_f under these conditions is in the form of a vee, Fig. 6.14 and, for various fixed loads, the curves are called the Vee-curves. A corresponding plot of p.f. variation is shown. Note that an over-excited synchronous motor takes a leading p.f. current. Used in this role, the machine is termed a 'synchronous capacitor' and is used for power factor correction purposes. Fig. 6.14 applies to the motoring case.

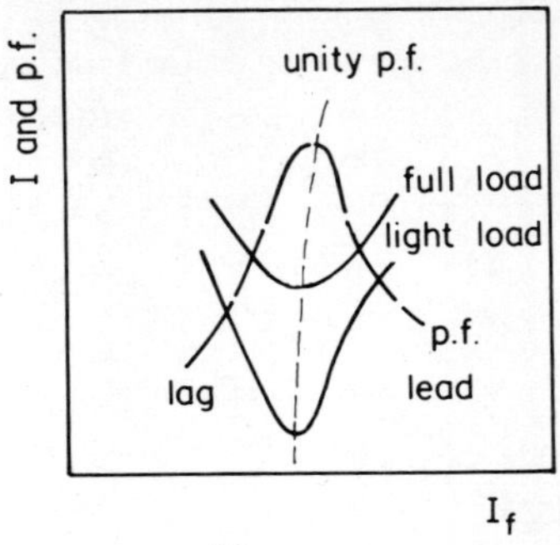

Fig. 6.14.

Now compare the impedance volt-drop triangle at unity p.f. with that for current at phase angle ϕ. Then

$$I_0 Z_s = (I \cos \phi) Z_s \propto (I \cos \phi) V$$
$$= \text{defined constant power.}$$

Hence (*i*) the locus of E for constant power, variable excitation is a straight line perpendicular to $I_0 Z_s$, or in general terms, a line at $\lambda = \tan^{-1} X_s/R$, to the horizontal.
(*ii*) a power scale may be marked on the diagram.

Considerations (a) and (b) are combined to provide an Electrical Load Diagram or Operating Chart for the machine performance, Fig. 6.15.

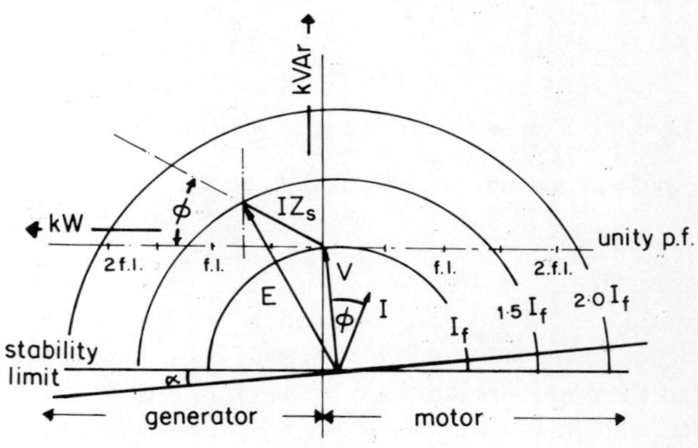

Fig. 6.15.

(c) The 'circle diagram' of current (synchronous motor case).

Consider the case of constant excitation, variable power.

A constant value of synchronous impedance Z_s is assumed. The phasors V, E and IZ_s then form the sides of a triangle, Fig. 6.16, from which it is seen :

$$(IZ_s)^2 = V^2 + E^2 - 2VE\cos\delta.$$

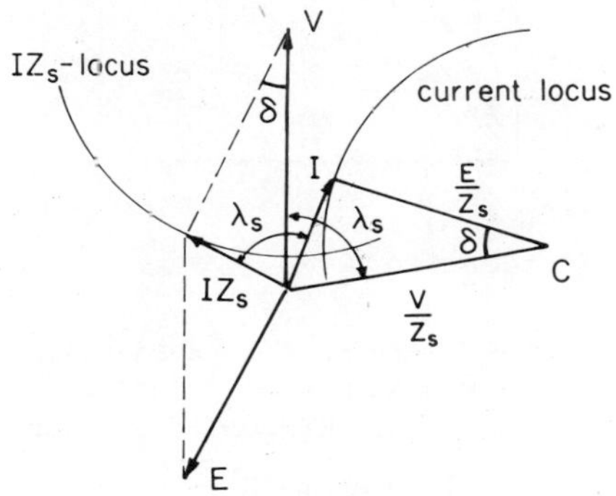

Fig. 6.16.

Since E is constant, the locus of IZ_s is a circle with centre at V and of radius $= E$. With constant Z_s, the locus of current I is another circle and I will be one side of a triangle, the other sides being V/Z_s and E/Z_s with angle δ between them. Alternatively, expressing impedance $= Z\,\epsilon^{j\lambda_s}$, the current locus is determined by changing the scale for voltages by $1/Z_s$ for currents, and rotating the voltage triangle by λ_s clockwise. The current locus will have radius $= E/Z_s$ and the angle $\lambda_s = \text{arc tan } X_s/R_a$.

Review of locus diagram methods

Diagrams such as the 'electrical load diagram', demonstrate that interpretation of the algebraic formulae pertaining to machine performance, and the corresponding phasor diagrams can provide convenient and rapid solutions to problems of machine performance. The loci depend initially on certain discreet and acceptable assumptions and lead to a measure of approximation in the solutions obtained.

The initial assumption of constancy of X_s may be improved upon by allowing for saturation, Fig. 6.17. The adjustment improves the accuracy of prediction for cylindrical rotor machines.

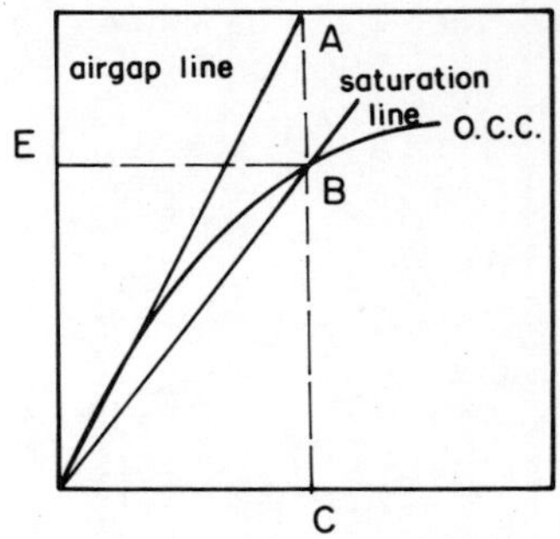

Saturation factor $K = \frac{BC}{AC}$

$$X_s = (X_1 + X_a) \text{ and } X_1 \text{ is assumed constant}$$

$$\text{sat.}X_a = K(X_s - X_1)$$

$$\text{sat.}X_s = X_1 + K(X_s - X_1) \qquad \text{(ref.2)}$$

Fig. 6.17.

But inaccuracy occurs when cylindrical rotor theory is applied to 'salient-pole' machines, particularly in respect of prediction of load angle and associated stability considerations.

The 'two-reaction' theory provides an improvement in the accuracy of prediction, and has led to the conception of two-axis principles which underlie the development of the 'generalised theory' of machines discussed in Chapter 8.

Practical steady-state stability limits

Consideration of the preceding phasor relationships has led to the development of a practical Power Chart for synchronous generators. In its simplest form, non-salient pole theory is assumed and resistance is neglected. Hence for a given load condition on a machine, a phasor diagram V, E and IX_s can be drawn, Fig. 6.18. Now it has been shown, (p. 172) that a power scale can be located as shown, ($\propto IX_s \cos\phi$). Similarly reactive-VA scales, (leading and lagging), are available. Since current and VA are proportional, (for constant V), appropriate scales can be determined as shown in the following example:

Machine specification: 415V, three-phase, star-connected stator, 12 kVA, 0·8 p.f., $X_s = 25\,\Omega$. Max. excitation = 2·5 p.u.

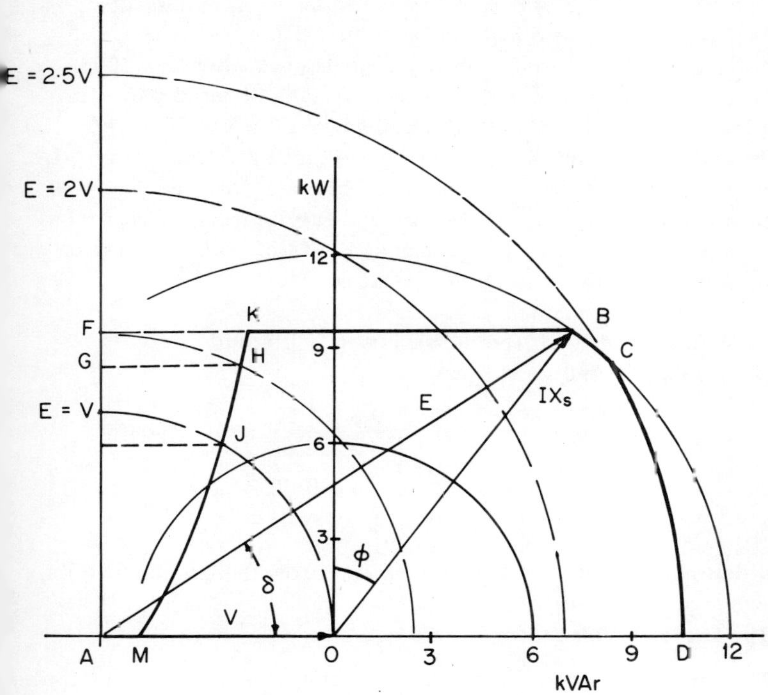

Fig. 6.18.

Let the voltage scale be 1 in. = 100 V/phase.

Then the current scale is 1in. $= \dfrac{\text{volts scale}}{X_s} = 4$A.

The VA scale/phase, 1 in. = (phase volts) × (current scale)
= 240 × 4 = 960 VA/ph.

For convenience, the VA scale is expressed in three-phase terms, viz:

1 in. = 3 × 0·96 = 2·88 kVA/three-phases.

Referring to Fig. 6.18, volts/phase = 240V = 2·4 in. = 0 A. Semi-circles about 0 represent various kVA-loadings. For the given machine the 12 kVA circle is one limit. Now the power rating, (at 0·8 p.f.) is 9·6 kW, limited by the prime-mover input. This limit is the constant-power line at 9·6 kW, and it meets the 12 kVA circle at B.

From C to D the output is limited by permissible rotor heating, viz., the excitation circle at 2·5 p.u.

Now the theoretical stability limit occurs when $\delta = 90°$, (at F). However a margin of some 10–20% of rated power is allowed before instability is reached, say 9·6 kW/10 = FG. But the operating point must be on the correct excitation circle, i.e. projected to H. Further points are obtained, such as J corresponding to p.u. excitation, and the limiting curve $KHJM$ is determined. Hence the practical limits of stable operation are determined, within the curve $MKBCD$.

Note: In this example:

At full rated load, (point B), $\cos\phi = 0{\cdot}8$ and E = 5·9 in. = 590 volts, (measured).

$$\therefore \quad \frac{240 \times 590}{25} \sin\delta = \frac{9{\cdot}6 \times 10^3}{3} \text{ watts/phase}$$

$$\therefore \quad \delta = \sin^{-1} 0{\cdot}562$$

$$= 34{\cdot}2°$$

which agrees with the measured value from the graphical form.

The two-reaction theory

The sinusoidal fundamental of the armature reaction when applie[d] to a uniform airgap produces a sinusoidal flux distribution. However, when applied to the distinctly non-uniform airgap of a salient-pole machine the flux distribution is of a different form. This problem is tackled by resolving the armature reaction into two components, one acting on the pole axis and termed the direct-axis reaction, the other acting on the interpolar gap and termed the quadrature-axis reaction.

Direct-axis m.m.f. is related to its flux by the direct-axis magnetization curve. Quadrature-axis m.m.f. acts on a relatively long airgap and may be assumed proportional to its flux.

A phasor diagram representing these conditions is developed as follows:

Working per phase, and with defined load conditions V, I and ϕ for a generator, the $I_a R$ and $I_a X_1$ voltage drops are added to determine airgap voltage E_1, (Fig. 6.19).

Now $\mathcal{F}_a$, due to I_a, is resolved into two component m.m.f's, $\mathcal{F}_d$ along the pole-axis and $\mathcal{F}_q$ on the quadrature-axis. $\mathcal{F}_d$ directly opposes $\mathcal{F}_t$, giving a resultant $\mathcal{F}$. $\mathcal{F}$ gives rise to a direct-axis flux which induces voltage E_q lagging upon it by 90. Component $\mathfrak{F}_q$ gives a quadrature-axis flux which induces E_d lagging $\mathfrak{F}_q$ by 90. These two voltages give a resultant E_1.

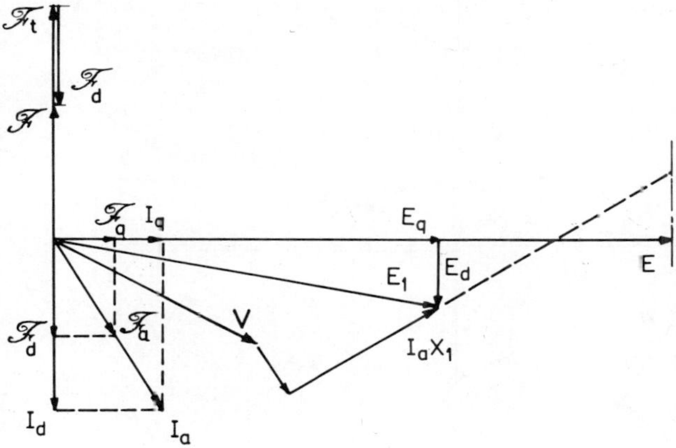

Fig. 6.19.

When the load is removed total main field m.m.f., $\mathfrak{F}_t$ is available to provide the open-circuit voltage E.

The concepts of $\mathfrak{F}_d$ and $\mathfrak{F}_q$ give rise to the notion of current components I_d and I_q resolved with respect to the flux axes. They must be distinguished clearly from the active and reactive components of current which are resolved with respect to V.

The effect of each component reaction is now considered in terms of a 'reactance equivalent to armature reaction' and designated X_{ad} and X_{aq} respectively. The phasor diagram is re-drawn, Fig. 6.20. Then by recognising $I_a X_1$ in terms of I_q and I_d we obtain the axis-reactances X_d and X_q where $X_d = X_{ad} + X_1$ and $X_q = X_{aq} + X_1$.

To construct the phasor diagram, load angle δ must be determined. Consider $I_a X_d$ drawn as $I_a X_s$ is drawn for uniform airgap theory, Fig. 6.20 (b). Then, by similar triangles, with the current triangle, the E phasor will intersect $I_a X_d$ at $I_a X_q$ and it follows that δ can be determined since I_a and X_q are known.

Note that the O.C. voltage E, on two-axis theory is not very different to E' obtained using X_s, at least in magnitude. But there is a significant discrepancy in the value of δ. If $X_d = X_q$ the two cases are identical, but in practice X_q/X_d for salient-pole machines is usually about 0·6.

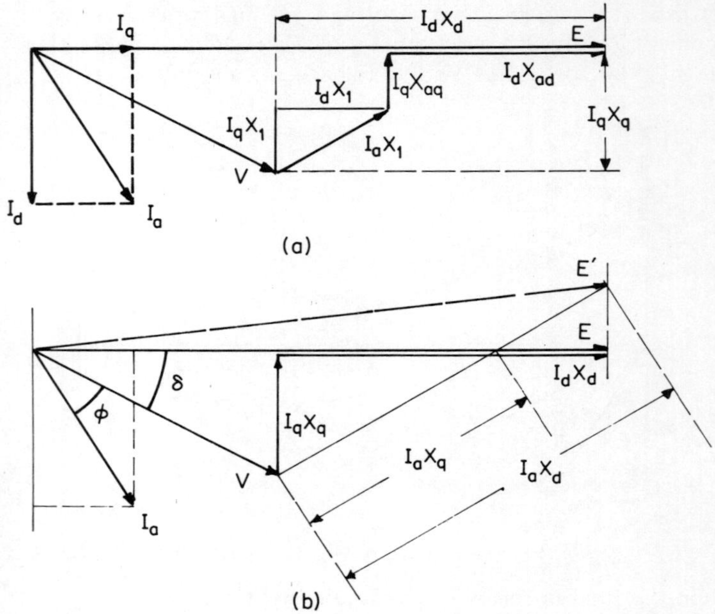

Fig. 6.20.

Power/angle relation, salient-pole case

Consider the two-axis phasor diagram, (neglecting resistance) Fig. 6.20.

$$I_q = I\cos(\phi + \delta);\; I_d = I\sin(\phi + \delta).$$

Then power/phase $= VI\cos\phi$

$$= V(I_q\cos\delta + I_d\sin\delta).$$

Now $V\cos\delta = E - I_d X_d \quad \therefore \quad I_d = \dfrac{E - V\cos\delta}{X_d}$

$V\sin\delta = I_q X_q \quad \therefore \quad I_q = \dfrac{V\sin\delta}{X_q}$

$$\text{Hence } P = V\left[\frac{V\sin\delta}{X_q}\cdot\cos\delta + \frac{E - V\cos\delta}{X_d}\cdot\sin\delta\right]$$

$$= \frac{VE}{X_d}\sin\delta + V^2\left(\frac{X_d - X_q}{2X_d X_q}\right)\sin 2\delta.$$

This relation is represented in Fig. 6.12. It will be noted that an additional term, $\propto \sin 2\delta$, occurs due to the development of reluctance torque in the salient-pole case. The result is that salient-pole machines are stiffer, i.e. the P/δ graph is steeper. Also maximum power occurs at a lesser angle, (about 60°).

Steady-state stability of salient-pole machines

In non-salient-pole theory it has been seen that the stability limit occurs when $\delta = 90°$. This occurs, (Fig. 6.12) when $\mathrm{d}P/\mathrm{d}\delta = 0$, or $(VE/X_s)\cos\delta = 0$. The slope of the P/δ curve therefore indicates the ability of a machine to 'contain' a change in load and is termed the Synchronising Power Coefficient. Also, the Synchronising Torque Coefficient $= 1/\omega_s\,(\mathrm{d}P/\mathrm{d}\delta)$, and in steady-state studies the assumption is that ω_s does not change due to ΔP.

For sudden changes in power, where the momentary change in ω_s is not negligible, oscillatory conditions can occur and a Transient Stability Limit must be envisaged. This is considered later.

In the present case, for a salient-pole machine:

$$\frac{\mathrm{d}P}{\mathrm{d}\delta} = \frac{VE}{X_d}\cos\delta + \frac{V^2(X_d - X_q)}{X_d X_q}\cos 2\delta$$

$$= \text{Synchronising Power Coefficient.}$$

Now, $\qquad \cos 2\delta = 2\cos^2\delta - 1$

$$\therefore \quad \frac{\mathrm{d}P}{\mathrm{d}\delta} = \frac{VE}{X_d}\cos\delta + \frac{2V^2(X_d - X_q)}{X_d X_q}\cos^2\delta - \frac{V^2(X_d - X_q)}{X_d X_q}.$$

Power is a maximum when $\mathrm{d}P/\mathrm{d}\delta = 0$. Then, multiplying across by $X_d X_q/2V^2(X_d - X_q)$:

$$\cos^2\delta + \frac{EX_q}{2V^2(X_d - X_q)}\cos\delta - \frac{1}{2} = 0$$

$$\therefore \quad \cos\delta_m = -\frac{EX_q}{4V(X_d - X_q)} \pm \frac{\sqrt{\left(\dfrac{E^2X_q^2}{4V^2(X_d - X_q)^2} + 2\right)}}{2}.$$

This may be written:

$$\cos \delta_m = \frac{1}{4}\left\{-\frac{EX_q}{V(X_d - X_q)} \pm \sqrt{\left(\left[\frac{EX_q}{V(X_d - X_q)}\right]^2 + 8\right)}\right\}.$$

Using this value for δ_m and by substitution in the equation for P, the load angle and power respectively at maximum power condition can be calculated.

Since the effects of saliency decrease as E increases, saliency is often neglected in steady-state stability studies provided $E > V$.

The Limaçon diagram of current

With the assumption of constant values for the machine parameters X_d and X_q and using conventional phasor methods of representation, it is shown that the phasors of current and generated voltage, in certain conditions of operation, follow loci which are mathematically known as limaçons. Mathematically the limaçon is a curve expressed by the polar equation, $r = (a \cos \theta + b)$. It may be defined as follows:

If $OA = a =$ the diameter of a circle, and OA is produced a distance 'b', then the locus of a point P, having polar coordinates (r, θ) with respect to OA as θ varies is a limaçon, when $OP = r = (a \cos \theta + b)$.

The proof is given here for the case of a synchronous motor operating on constant voltage bus-bars. The limaçon locus of current occurs for the conditions of variable loading at constant excitation. Resistance is neglected.

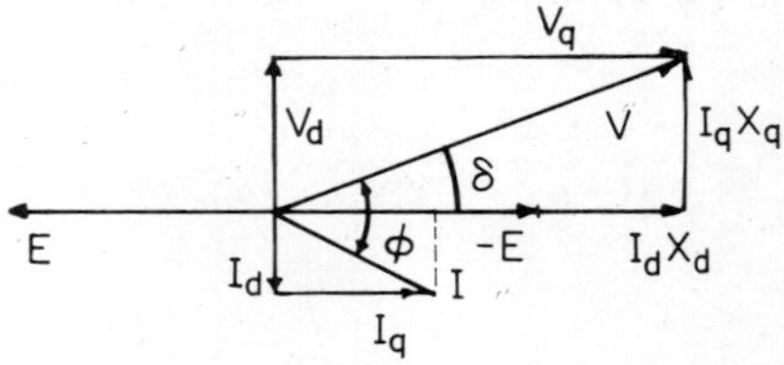

Fig. 6.21.

From the two-reaction phasor diagram, Fig. 6.21:

$$\mathbf{I} = \mathbf{I}_d + \mathbf{I}_q$$
$$\mathbf{V} = \mathbf{V}_d + \mathbf{V}_q$$

$$\mathbf{V}_d = -V\sin\delta\cdot\epsilon^{-j\delta} = \mathbf{I}_q(jX_q)$$

$$\mathbf{V}_q = jV\cos\delta\cdot\epsilon^{-j\delta} = -\mathbf{E} + \mathbf{I}_d(jX_d)$$

$$\therefore\quad \mathbf{I}_q = \frac{-V\sin\delta\cdot\epsilon^{-j\delta}}{jX_q} = j\,\frac{V\sin\delta}{X_q}\cdot\epsilon^{-j\delta}$$

$$\mathbf{I}_d = \frac{jV\cos\delta\cdot\epsilon^{-j\delta} + \mathbf{E}}{jX_d}$$

$$= \frac{V\cos\delta\cdot\epsilon^{-j\delta}}{X_d} - j\,\frac{\mathbf{E}}{X_d}.$$

Now

$$\mathbf{E} = E\underline{/-(\pi/2+\delta)}$$

$$-\mathbf{E} = E\underline{/\pi-(\pi/2+\delta)} = E\underline{/(\pi/2-\delta)}$$

$$-j\,\mathbf{E} = E\epsilon^{j(\pi-\delta)}$$

$$\therefore\quad \mathbf{I}_d = \frac{V\cos\delta\cdot\epsilon^{-j\delta}}{X_d} + \frac{E}{X_d}\cdot\epsilon^{j(\pi-\delta)}.$$

Substituting the expressions for $\mathbf{I}_d$ and $\mathbf{I}_q$ in $\mathbf{I}$:

$$\mathbf{I} = jV\cdot\frac{\sin\delta}{X_q}\cdot\epsilon^{-j\delta} + V\frac{\cos\delta}{X_d}\cdot\epsilon^{-j\delta} + \frac{E}{X_d}\,\epsilon^{j(\pi-\delta)}.$$

Now

$$\sin\delta = \frac{1}{2j}(\epsilon^{j\delta} - \epsilon^{-j\delta})$$

$$\cos\delta = \frac{1}{2}(\epsilon^{j\delta} + \epsilon^{-j\delta})$$

$$\therefore\ \mathbf{I} = j\,\frac{V}{X_q}\,\epsilon^{-j\delta}\,\frac{1}{2j}(\epsilon^{j\delta} - \epsilon^{-j\delta}) + \frac{V}{X_d}\,\epsilon^{-j\delta}\,\frac{1}{2}(\epsilon^{j\delta} + \epsilon^{-j\delta}) + \frac{E}{X_d}\,\epsilon^{j(\pi-\delta)}$$

$$= \frac{V}{2X_q}(1 - \epsilon^{-j2\delta}) + \frac{V}{2X_d}(1 + \epsilon^{-j2\delta}) + \frac{E}{X_d}\,\epsilon^{j(\pi-\delta)}$$

$$= \frac{V}{2}\left(\frac{1}{X_q} + \frac{1}{X_d}\right) + \frac{V}{2}\left(\frac{1}{X_q} - \frac{1}{X_d}\right)\epsilon^{j(\pi-2\delta)} + \frac{E}{X_d}\,\epsilon^{j(\pi-\delta)}$$

$$= I_1 + I_2\,\epsilon^{j(\pi-2\delta)} + I_3\,\epsilon^{j(\pi-\delta)}.$$

Thus, the current phasor is determined by three component phasors, as shown graphically in Fig. 6.22.

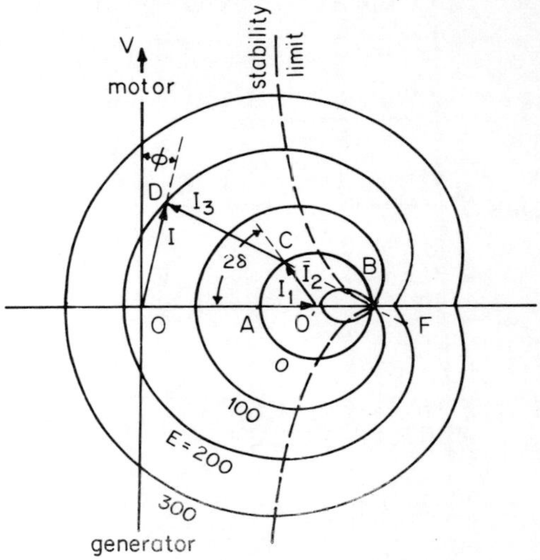

Fig. 6.22.

The solution is obtained graphically as follows, Fig. 6.22.

Mark off $$OO' = I_1 = \frac{V}{2}\left(\frac{1}{X_q} + \frac{1}{X_d}\right).$$

With centre O′ draw a circle of radius

$$= I_2 = \frac{V}{2}\left(\frac{1}{X_q} - \frac{1}{X_d}\right).$$

Set off I_2 = O′C at an angle 2δ to OO′. Join BC and produce to make CD = I_3 = E/X_d. Note that the angle DBO = δ. The phasor OD determines the current I in magnitude and phase.

Note that $$BC = BA\cos\delta = V\left(\frac{1}{X_q} - \frac{1}{X_d}\right)\cos\delta$$

and $$CD = \frac{E}{X_d}$$

$$\therefore \quad BD = \frac{E}{X_d} + V\left(\frac{1}{X_q} - \frac{1}{X_d}\right)\cos\delta$$

= the equation of a limaçon as δ varies, when E, V, X_d and X_q are all constants.

This solution illustrates several important characteristics of salient-pole machine performance. Note that the component current, $I_2\,\epsilon^{j(\pi-2\delta)}$, demonstrates the influence of saliency, which is the superimposition upon the fundamental torque of a second-harmonic constant-amplitude torque. The fundamental torque varies with excitation, viz:

$$\frac{E}{X_d}\,\epsilon^{j(\pi-\delta)}.$$

The circle on AB as diameter is the locus of current when $E = 0$, i.e. for the machine when running as a reluctance machine. Operation of the machine with reversed or 'negative' excitation is determined by the limaçon, viz. CF = CD; it is of interest to note that part of the re-entrant loop occurs within the region of stable operation. The locus of maximum ($I \cos \phi$) is the theoretical stability limit, whilst horizontal lines drawn on the diagram are constant-power loci of current, and constant power lines when V is constant.

The generator, or 'mirror image' diagram has been developed as a practical power chart for salient-pole generators (Ref. 6), similar to that developed for the non-salient pole case, (see also Ref. 7).

Measurement of reactances and reaction

Under s.c. conditions, a synchronous machine operates at virtually zero p.f. lagging conditions, since the machine circuit is largely reactive. The phasor diagram under these conditions is shown in Fig. 6.23. The internal volt drop may be assumed to be due to:

(i) a synchronous impedance, Z_s or
(ii) total direct demagnetization by $\mathcal{F}_a$ or
(iii) a combination of $\mathcal{F}_a$ and a small X_1.

An estimate of the individual magnitudes is made using the open and short- circuit characteristics of the machine. For both characteristics the machine is driven at synchronous speed. Observations are taken of generated volts/field current for the O.C.C. and of armature current/field current for the S.C.C., Fig. 6.24.

Now at I_{FL} on s.c., 0A field amps are required. On o.c., 0A field amps give a generated voltage E. Then using assumption (*i*):

$$\text{synchronous impedance, } Z_s = E/I_{FL}.$$

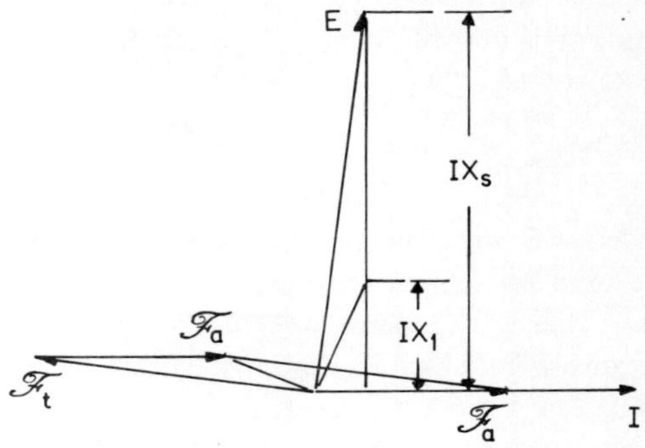

Fig. 6.23.

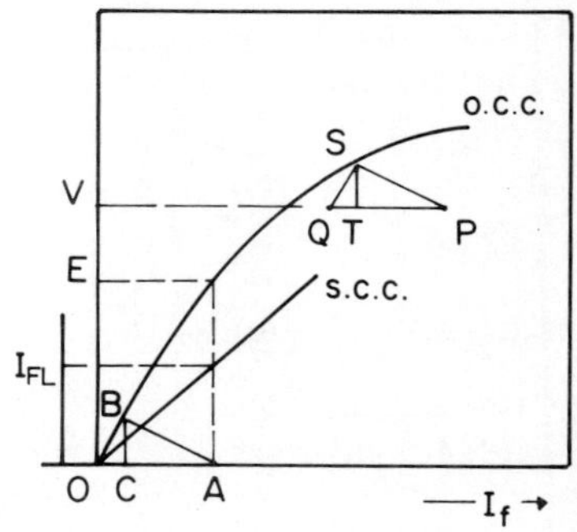

Fig. 6.24.

In general, $$Z_s = \frac{E_{oc}}{I_{sc}}$$

where E_{oc} and I_{sc} are taken at the same value of field current. Allowing for resistance :

$$\text{synchronous reactance, } X_s = \sqrt{(Z_s^2 - R^2)}.$$

If assumption (*ii*) is adopted :

$$0A = \mathcal{F}_a .$$

When assumption (*iii*) is used, only a part of 0A, say CA $= \mathcal{F}_a$ and BC $= IX_1$. It is necessary to determine the point C

experimentally and the procedure is termed Potier's Method. Now $\mathcal{F}_a$ and IX_1 may be assumed constant for a given current. Hence the triangle ABC will be constant for any load at I_{FL}, zero p.f. lagging. An additional test point is obtained, with a loading at I_{FL} zero p.f. lagging and generally at rated voltage V. Commonly an under-excited synchronous motor is used to provide this loading. This determines the point P, termed the Potier Point.

By similar triangles PQ = OA and QS//OB and the apex S of the triangle is determined. Hence ST = IX_p and TP = $\mathcal{F}_a$, where X_p is termed the Potier reactance. X_p is somewhat greater than X_1 due to increased pole and yoke flux leakage on load, but X_p obtained at high values of saturation gives a good approximation to X_1.

The two axis-reactances are determined as follows:

(i) The slip test

A low voltage balanced three-phase supply of rated frequency is applied to the three-phase stator, thereby producing a synchronous rotating field. The rotor is now driven at near-synchronous speed, with the rotor winding open-circuited. The airgap reluctance varies at the small slip frequency and stator voltmeter and ammeter readings fluctuate. From the amplitudes of the swings, per phase, or by oscillogram:

$$X_d = \frac{V_{max}}{I_{min}} \quad \text{and} \quad X_q = \frac{V_{min}}{I_{max}}.$$

These correspond to the minimum and maximum of the airgap reluctance, which occur on the direct and quadrature magnetic axes of the machine. Different values of slip are used until consistent results are obtained.

Since the short-circuit condition under which X_s is obtained corresponds to direct-axis demagnetization, then $X_d = X_s$. The American Standards Association recommend that only the ratio X_q/X_d be determined by the slip test and since $X_d = X_s$, X_q is determined by ratio.

(ii) The pull-out or maximum-lagging-current test

The machine is run normally as a synchronous motor, unloaded. By reducing the field current to zero and then slowly increasing the field current in the reversed polarity until the machine pulls out of synchronism, instantaneously $\delta = 90°$ elect., and:

$$\frac{V_{ph}}{I_{po}} = X_q$$

where I_{po} is the maximum stable current/phase before the machine 'slips a pole'.

Synchronising three-phase a.c. supplies

When certain conditions have been obtained, the connection of one three-phase supply in parallel with a second three-phase supply is termed 'synchronising'. In general, the term implies that the necessary conditions have been satisfied. The necessary conditions are that the two three-phase supplies:

1. have the same phase rotation,
2. have the same voltage,
3. have the same frequency,
4. are 'in synchronism' at the instant of closing the switch.

Item 1 is checked by a test such as the Varley phase-rotation test. This is normally done, once and for all, when a machine is installed, in keeping with the normal direction of rotation, and polarity of field excitation supply. Items 2 and 3 are adjusted, generally at the incoming machine, at the time switching is required. Finally item 4 is checked by means of a 'synchroscope'. A synchroscope, in this context, indicates whether or not two a.c. supplies are instantaneously in phase.

The Varley phase-rotation test

An unbalanced star-connected load comprising two equal non-inductive resistances, (usually lamps), and a reactance, Fig. 6.25. is connected to the three-phase terminals to be tested. Owing to the absence of a neutral wire the currents must balance at the star-point S, i.e. the current in the reactance must balance the resultant of the lamp currents. The lamp currents will be proportional to, and in phase with the phase voltages V_1 and V_2, Fig. 6.26. The phasor resultant of the currents I_1 and I_2 must pass through P, the mid-point of BC. Thus P is a fixed point

and I_L represents the current in the reactance, (assuming inductance). Then I_L lags V_L by 90° and ∠ASP = 90°. Hence the locus of S is a circle on AP as diameter.

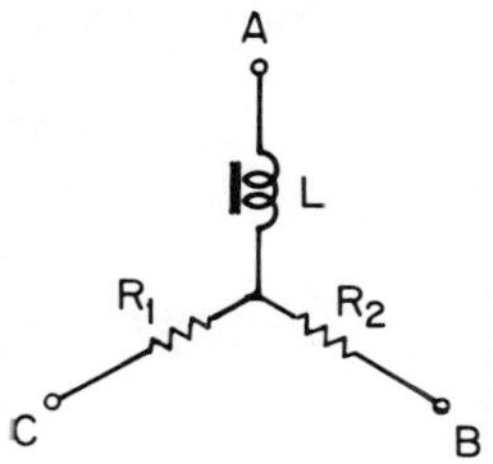

Fig. 6.25.

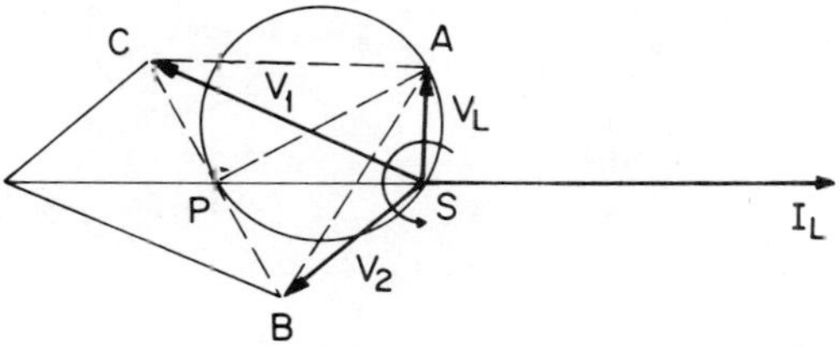

Fig. 6.26.

With inductive reactance, lamp SB has a lower voltage than SC and so is dimmer and indicates the leading phase, giving sequence BCA.

With capacitative reactance the star point S swings to the other side of AP and the brighter lamp indicates the leading phase.

Note that you are the stationary observer and the phasors pass anticlockwise and this is the sequence.

Lamp synchroscopes

(*i*) Three high resistance lamps connected across the parallelling switch, Fig. 6.27 (a), can be used as a synchroscope. The phasor diagram, Fig. 6.27 (b) shows that, when the supplies are in phase the three lamps are extinguished and the arrangement is termed 'synchronizing with lamps dark'.

(*ii*) By cross-connecting lamps 2 and 3, Fig. 6.28 (a), the correct instant for synchronizing occurs when lamps 2 and 3 are of equal brilliance and lamp 1 is extinguished. The arrangement is termed 'synchronizing with lamps bright', since often only the cross-connected lamps are displayed on the switchboard. It is known as the 'Siemens-Halske' method and gives additional information, since the sequence in which the lamps light indicates if the incoming machine is too fast or too slow.

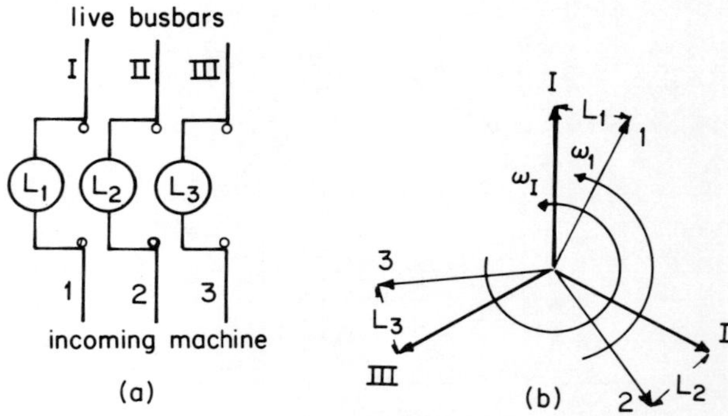

Fig. 6.27.

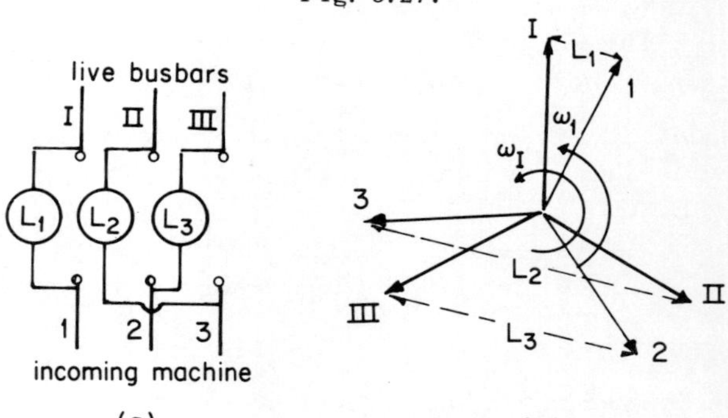

Fig. 6.28.

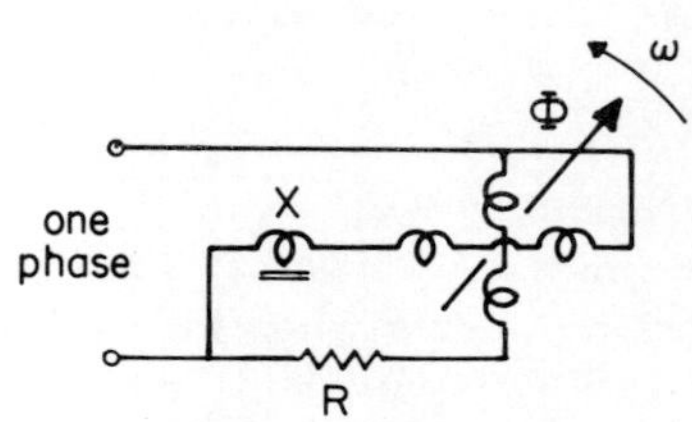

Fig. 6.29.

Instrument synchroscopes

In general these operate on induction motor principles. The rotor and stator elements are each fed from corresponding single phases of the supplies being synchronized. Each element has two coils physically at 90° and fed by a 'split-phase' arrangement, Fig. 6.29. When 'in phase' conditions obtain, no misalignment occurs between rotor and stator fluxes and no torque occurs. Indication is given by a pointer on a dial.

Transient performance of synchronous machines

Transient conditions occur in the performance of synchronous machines due to an abrupt change in the electrical or the mechanical loading. Accurate measurement of the time constants involved may be made from analysis of oscillograms obtained by direct test. Complete theoretical analysis involves a complicated electromagnetic situation and difficult mathematics. However, an approximate analysis can be developed which leads to an understanding of the principal factors which control the transient behaviour. This enables approximate predictions of transient performance to be made and gives guidance in the choice of suitable circuit-breakers and other protective devices. More exact analyses provide the machine designer with important criteria to guide a design.

The electrical time constants are usually much shorter than the mechanical time constants and the study is made in two parts, viz; (a) concerning transient electrical phenomena, (b) concerning electromechanical dynamics.

(a) The sudden three-phase short-circuit

Consider a three-phase synchronous generator operating normally on open circuit. Sudden application of a zero-resistance s/c to the terminals may be thought of in terms of a step-function of voltage, $V \sin \omega t$, applied to a phase winding of effective inductance L' and resistance R. Then, as shown on p. 19:

$$i = I_{max} [\sin(\omega t - \phi) - \sin(\omega t' - \phi) \epsilon^{(t'-t)/T'}]$$

where the instant of s/c occurs t' sec after zero in the voltage cycle. It is assumed that no change in speed occurs during the transient period. Also, $\tan \phi = \omega L'/R$ and transient time-constant $T' = L'/R$.

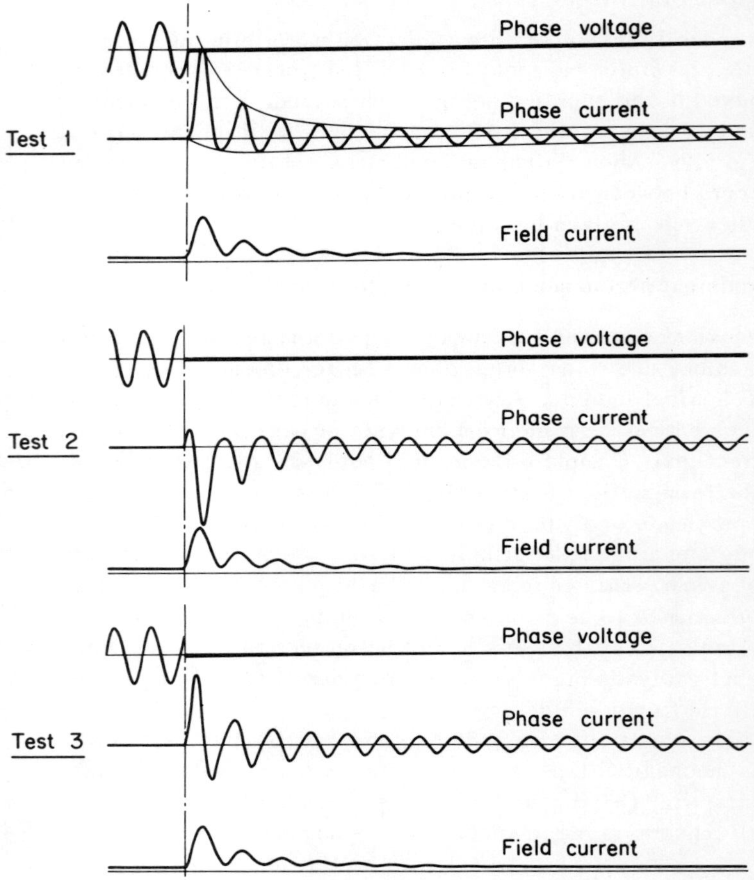

Fig. 6.30.

The true value of L' is complicated, but is here thought of in terms of an effective value, $L' = (1 - k^2)\,L = \sigma L$ and σ = stator/rotor leakage factor. This is reconsidered later.

A further approximation may be made, since R is small and $R \ll \omega L'$ and hence $\phi \to 90°$, then:

$$i = I_{max}\,[\cos \omega t - (\cos \omega t')\,\epsilon^{(t'-t)/T'}]$$

$$= \text{steady-state part} + \text{transient (d.c.) part.}$$

It is valuable at this stage to examine an actual s/c oscillogram Figure 6.30 shows oscillograms obtained, from the same phase,

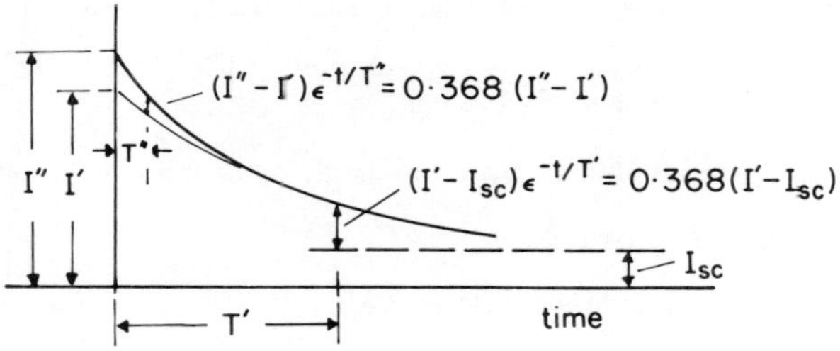

(a)

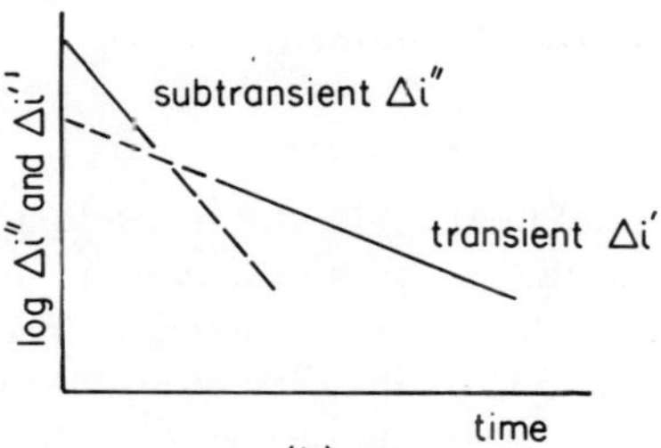

(b)

Fig. 6.31.

during sudden three-phase symmetrical s/c tests on a small synchronous generator, the s/c being applied on three different occasions. The presence of asymmetry due to the d.c. or transient part will be noted. This can be removed graphically. The resulting symmetrical a.c. component is now plotted, in terms of r.m.s. values, on linear and log-linear graph paper. The results exhibit important characteristics, viz: (Fig. 6.31):

(*i*) Three distinct stages are evident, namely subtransient, transient and final steady-state. The time constant appropriate to the exponential governing each stage is measured as shown in Fig. 6.31, (i.e. when $\epsilon^{-t/T} = 0{\cdot}368,\ t = T$).

(*ii*) The initial r.m.s. current corresponding to each stage can be defined, in terms of initial o/c voltage, E_0.

namely: $$I'' = \frac{E_0}{X''},\quad I' = \frac{E_0}{X'} \text{ and } I_{sc} = \frac{E_0}{X_d}$$

with the definitions:

X_d = synchronous reactance, (steady-state),
X' = transient reactance,
X'' = subtransient reactance.

Hence the r.m.s. value of the symmetrical a.c. current is expressed:

$$i_{ac} = (I'' - I')\,\epsilon^{-t/T_d''} + (I' - I_{sc})\,\epsilon^{-t/T_d'} + I_{sc}\,.$$

The d.c. component may be measured, or calculated from:

$$i_{dc} = \sqrt{2}\,I''(\cos\omega t')\,\epsilon^{-(t-t')/T_a}.$$

Then, the total s/c current, r.m.s. value:

$$i_{\mathrm{rms}} = (i_{ac}^2 + i_{dc}^2)^{\frac{1}{2}}.$$

The following physical interpretation enables the constants involved to be more exactly defined.

Approximate physical interpretation

With the assumption of negligible stator winding resistance, under s/c conditions the stator m.m.f. wave is directly demagnetizing, i.e. on the direct, (*d*-axis) or field axis of the machine. Now the total flux-linkages of a closed circuit must remain constant, (viz. 'principle of constant flux linkages', – because if a change did occur the additional e.m.f. induced would contradict Kirchhoff's voltage law). Hence the rapid change in stator current must be counterbalanced by an induced current in the field circuit. The decay of these transient phenomena is governed by the resistances and reactances, i.e. time constants of the associated circuits.

The transient change in field current, i.e. excitation, largely determines the behaviour of the stator current. In addition, closed damper circuits, which have much shorter time constants, provide the initial subtransient increase in stator current.

As seen in the transient analysis of the simple RL circuit, the s/c current contains a.c. and d.c. transient components, depending on the instant of s/c. The d.c. component in the stator dies away dependant on the stator resistance and inductance, (T_a). Relative to the synchronously rotating rotor, this stator d.c. induces a fundamental-frequency component in the field circuit. The flux due to this single-phase induced a.c. field current can be resolved into counter-rotating (conjugate) fluxes, one which counterbalances the stator d.c. component, the other affects the stator at twice synchronous frequency and induces in the stator a 2nd harmonic (commonly neglected in analyses), and can generally be noticed in experimental oscillogr

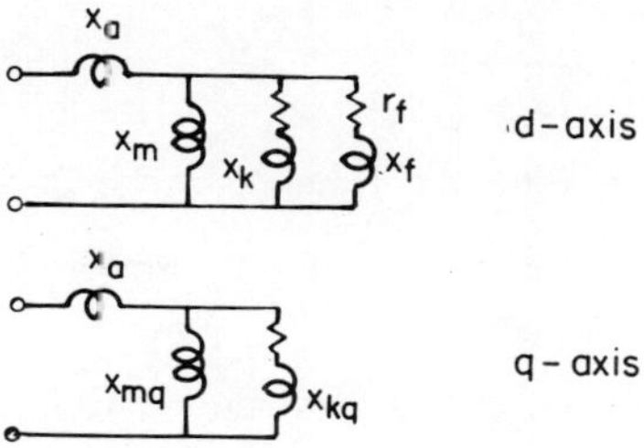

Fig. 6.32.

The nature of the time constants can be deduced by considering an elementary equivalent circuit, Fig. 6.32. The notation used is conventional, with the following definitions :

r_a = armature resistance,

r_f = field resistance,

x_a = armature leakage reactance,

x_f = field leakage reactance,

x_m = direct-axis magnetizing reactance.

Referring to Fig. 6.32:

On steady-state: $x_d = x_a + x_m$.

On short-circuit: $x'_d = x_a + \dfrac{x_f\, x_m}{x_f + x_m}$,

that is, x'_d is the armature leakage reactance plus a modified field-leakage reactance. The armature short-circuit current, (r.m.s.):

$$I' = \frac{E_0}{x'_d}.$$

The direct-axis open-circuit transient time constant is determined by x_f in series with x_m, thus :

$$T'_{do} = \frac{1}{\omega r_f}(x_f + x_m).$$

Under s/c conditions the effective field reactance affecting the decay of the field transient is x_f, in series with x_m and x_a in parallel.

$$\therefore \quad x_f' = x_f + \frac{x_a x_m}{x_a + x_m} = \frac{(x_f + x_m)}{(x_a + x_m)}\left(x_a + \frac{x_m x_f}{x_m + x_f}\right).$$

The direct-axis transient s/c time constant is thus:

$$T_d' = \frac{(x_f + x_m)}{\omega r_f}\frac{x_d'}{x_d} = T_{do}' \cdot \frac{x_d'}{x_d}.$$

With the introduction of a s/c damper winding, (suffix k), and a similar reasoning, the subtransient time constants are:

$$T_d'' = \frac{1}{\omega r_k}\left(x_k + \frac{x_m x_a x_f}{x_m x_a + x_m x_f + x_a x_f}\right)$$

and $$T_{do}'' = \frac{1}{\omega r_k}\left(x_k + \frac{x_m x_f}{x_m + x_f}\right).$$

In interpreting T_a, the time constant which applies to the decay of the armature d.c. component, it is recognised that this component of m.m.f. 'sees' a rotor which is rotating synchronously. Thus the leakage flux path varies between x_d'' and x_q'', and is commonly taken as the mean, viz:

$$T_a = \frac{1}{\omega r_a}\left(\frac{x_d'' + x_q''}{2}\right).$$

Note that the q-axis does not carry a field winding, and thus:

$$x_q = (x_a + x_{mq}) \text{ and } x_q'' = \left(x_a + \frac{x_{mq} x_{kq}}{x_{mq} + x_{kq}}\right).$$

(*b*) *Electromechanical stability of synchronous machines*

The stability of a synchronous machine is its ability to return to normal synchronous conditions following a disturbance due to change of input or output. Change of input can arise due to a pulsating drive from a reciprocating prime mover, and may lead to a condition termed 'cyclic irregularity'. An undesirable 'hunt' may develop in the system if a resonant or under-damped condition prevails. Transient conditions arise when the disturbanc is abrupt, e.g. sudden fault or load-switching on a generator, and a synchronising torque is developed which tends to restore stable synchronous conditions.

The dynamical behaviour of the machine is governed by the inertia of the rotating mass, the damping and by the stiffness of the machine and associated network. The natural response may be defined in terms of a differential equation. The forcing function may be periodic or a step function.

Basic dynamics. Consider the case for a generator. We have, equating torques:

$$Ⓣ = J\,\mathrm{d}\omega/\mathrm{d}t + D\omega + Ⓣ_e$$

where Ⓣ = mechanical shaft torque, J = moment of inertia of the rotating mass in kilogram-metres^2, D = damping, $\propto$ velocity and $Ⓣ_e$ = electromagnetic or synchronising torque, determined in the torque/angle relation.

The kinetic energy absorbed by a rotating mass = $\frac{1}{2}J\omega^2$ joules, and the angular momentum $M = J\omega$ joules.sec/rad. For large machines an inertia constant H is defined as the stored energy per VA rating. Thus for a machine of $G\ MVA$:

$$GH = \text{stored energy in megajoules,}$$

$$= \tfrac{1}{2}J\omega^2 \times 10^{-6} = \tfrac{1}{2}M\omega \times 10^{-6} \text{ megajoules.}$$

Angular displacement from the synchronous position for electrical machines is commonly expressed in terms of an electrical angle in electrical degrees, and $\delta_{\text{elect}} = p\delta_{\text{mech}}$. Thus for a sudden departure from synchronous position by $\delta°$ elect, we can say:

$$\text{Inertia Torque, } Ⓣ_i = M\,\mathrm{d}^2\delta/\mathrm{d}t^2$$

when M has been adjusted for the new units. Then inertia power may be expressed

$$= \omega Ⓣ_i = J \times \frac{2\pi N}{60} \times \frac{\pi}{180p}\,(\mathrm{d}^2\delta/\mathrm{d}t^2)$$

where N = synchronous speed in r.p.m., and the departure from N is considered negligible. Note that ω = synchronous speed in mech. rad/sec, ω_s = synchronous speed in elect. rad/sec = $p\omega$.

Then $M = 0{\cdot}001825\,JN/p$ where p = pairs of poles,

$= 0{\cdot}11\,Jf/p^2$ where f = supply frequency, c/s,

and M is in joules.sec/elect.deg = watts/elect.deg/sec^2. If J is expressed in lb. ft^2, (Wk^2), then $M = 0{\cdot}0046\ \mathrm{Wk^2f/p^2}$.

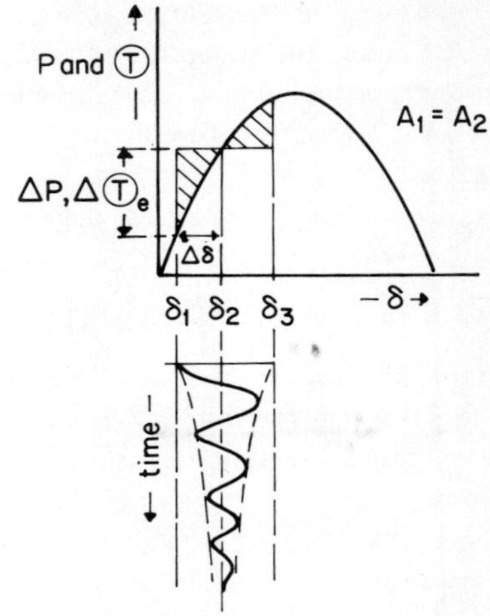

Fig. 6.33.

Now, for a sudden change of load, $\Delta Ⓣ_e$, the rotor angle is initially maintained constant at δ_1 by the inertia of the drive, and kinetic energy is extracted from the stored energy in the rotating system. Then the speed drops by $(\omega_s - \omega_r)$ elect.rads/sec. This can be expressed $= \Delta\delta/\Delta t$ in elect.deg/sec.

In the limit as $\Delta t \to 0$, $\Delta\delta/\Delta t = \mathrm{d}\delta/\mathrm{d}t$ and δ_1 swings to δ_2, (and beyond ?), Fig. 6.33. Then :

$$\Delta Ⓣ = J\, \mathrm{d}^2\delta/\mathrm{d}t^2 + D\, \mathrm{d}\delta/\mathrm{d}t + \Delta Ⓣ_e.$$

Since $\mathrm{d}\delta/\mathrm{d}t \ll \omega_s'$, where $\omega_s' = (180/\pi)\,\omega_s$ elect.deg/sec

$$\therefore \quad \Delta P = \omega_s' \left(J\, \mathrm{d}^2\delta/\mathrm{d}t^2 + D\, \mathrm{d}\delta/\mathrm{d}t + \Delta Ⓣ_e\right)$$

where $\omega_s' J$ = angular momentum, M, in joules.sec/elect.deg.,

$\omega_s' D = P_d$ = damping in watts/elect.deg/sec.,

$\omega_s' \Delta Ⓣ_e = \Delta P_s$, the synchronising power and is a function of δ.

Hence we can write :

$$\Delta P = M\, \mathrm{d}^2\delta/\mathrm{d}t^2 + P_d\, \mathrm{d}\delta/\mathrm{d}t + \Delta P_s\,.$$

If damping is neglected, the system is oscillatory, and :

$$M\, \mathrm{d}^2\delta/\mathrm{d}t^2 = \Delta P_s\,.$$

Multiplying across by $\mathrm{d}\delta/\mathrm{d}t$:

$$M\,\mathrm{d}^2\delta/\mathrm{d}t^2\,(\mathrm{d}\delta/\mathrm{d}t) = \tfrac{1}{2}M\,\frac{\mathrm{d}}{\mathrm{d}t}(\mathrm{d}\delta/\mathrm{d}t)^2$$

$$= \Delta P_s\,(\mathrm{d}\delta/\mathrm{d}t).$$

Integrating both sides and re-arranging:

$$\frac{\mathrm{d}\delta}{\mathrm{d}t} = \sqrt{\left[\frac{2}{M}\int_{\delta_1}^{\delta} \Delta P_s \cdot \mathrm{d}\delta\right]}.$$

The limits of 'swing' occur when $\mathrm{d}\delta/\mathrm{d}t = 0$, i.e., at δ_1 and δ_3 and this implies that

$$\int_{\delta_1}^{\delta_3} \Delta P_s \cdot \mathrm{d}\delta = 0.$$

This can be written:

$$\int_{\delta_1}^{\delta_2} \Delta P_s \cdot \mathrm{d}\delta + \int_{\delta_2}^{\delta_3} \Delta P_s \cdot \mathrm{d}\delta = 0.$$

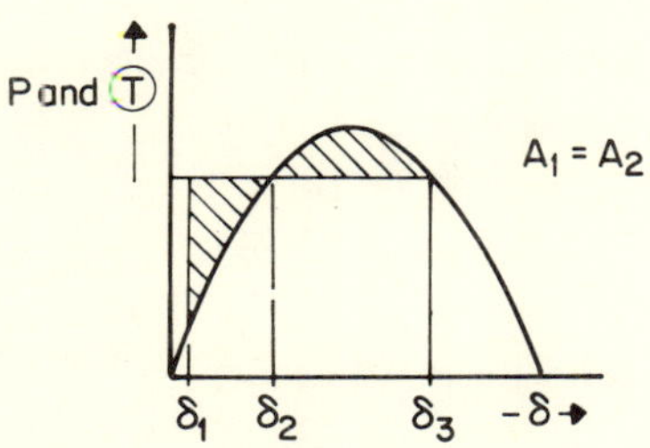

Fig. 6.34.

Referring to the shaded areas in Fig. 6.33, this means that $A_1 + A_2 = 0$, or $A_1 = A_2$ in magnitude. This is known as the Equal-Area Criterion, and gives the maximum swing for stable operation.

Critical conditions are reached when $\delta_2 = (180° - \delta_3)$, Fig. 6.34. This restricts the load to which the machine may be increased from an initial loading, without losing synchronism. In this case, δ_2 is termed the 'critical rotor angle'. The time corresponding to this critical angle is of great importance in relation to the time-setting of protective relays or circuit breakers (Ref. 4).

Small oscillations. The equation of motion considered above is non-linear since the synchronising power is a function of $\sin\delta$. Where the value of δ is small, ($< 30°$), then a linear relationship may be assumed for synchronising power, proportional to δ. We can then write the equation of motion:

$$M\,\mathrm{d}^2\delta/\mathrm{d}t^2 + P_d\,\mathrm{d}\delta/\mathrm{d}t + P_s\,\delta = P$$

i.e. as a linear differential equation and P_s is the slope of the P/δ graph. The solution of the equation gives:

$$\text{Steady-state, } \delta_{ss} = P/P_s$$

$$\text{Natural 'frequency', } \omega_n = (P_s/M)^{\frac{1}{2}} \text{ and } f_n = \frac{1}{2\pi}(P_s/M)^{\frac{1}{2}}$$

$$\text{Also, since } P_s = \mathrm{d}P/\mathrm{d}\delta = P_m \cos\delta$$

$$\therefore\ f_n = \frac{1}{2\pi}\sqrt{\left(\frac{mVE}{MX_s}\cos\delta\right)}$$

where m = phases, M applies to the whole machine; V, E and X_s are normally per phase.

$$\text{The damping ratio, } \zeta = \frac{P_d}{2\sqrt{(MP_s)}}.$$

With $\zeta = 1$, critical damping occurs.

When $\zeta < 1$, the solution is:

$$\delta = \delta_{ss}\left[1 - \frac{1}{(1-\zeta^2)^{\frac{1}{2}}}\,\epsilon^{-\zeta\omega_n t}\sin\{(1-\zeta^2)^{\frac{1}{2}}\,\omega_n t + \phi\}\right]$$

and with damping: $\omega_d = \omega_n(1-\zeta^2)^{\frac{1}{2}}$

and $$\tan\phi = \frac{(1-\zeta^2)^{\frac{1}{2}}}{\zeta}.$$

REFERENCES

1. SAY, M.G. (1961). *Performance and Design of a.c. Machines*, Pitmans,
2. FITZGERALD, A.E. and KINGSLEY, C. (1961). *Electric Machinery*, 2nd ed., McGraw-Hill, New York.
3. HINDMARSH, J. (1965). *Electrical Machines*, Pergamon Press, Oxford.
4. WEEDY, B.M. (1967). *Electric Power Systems*, Wiley, New York.
5. LANGSDORF, A. (1937). *Contributions to Synchronous Machine Theory*, Trans. A.I.E.E. 56. p. 41; also discussion p. 907.
6. WALKER, J.H. (1953). *Operating Characteristics of Salient-Pole Machines*, Proc. I.E.E. 100, Part II, p. 13.
7. SZWANDER, W. (1944). *Fundamental Characteristics of Synchronous Turbogenerators*, J.I.E.E. 91, Part II, p. 185.

EXAMPLES

EXAMPLE 6.1.

A three-phase star-connected synchronous generator delivers 600 A at unity power factor to constant voltage 11 kV bus-bars. The machine stator has impedance $(0{\cdot}5 + j8)\,\Omega$/phase. Calculate, (a) the load angle, (b) the new load angle, the current and the power factor when the excitation is raised by 20% and the power output remains unchanged, (c) the maximum power output at the new excitation.

Solution:

A graphical solution may be used, Fig. Ex. 6.1 and this is described below. With an analytical solution, a sketch of the phasor diagram is essential, not necessarily to scale.

Analytical: Phase volts $= \dfrac{11000}{\sqrt{3}} = 6356\text{V}$

$$I_p = I_L = 600\text{A at u.p.f.}$$

(a) $\therefore\ E^2 = (V + IR)^2 + (IX)^2$ Fig. Ex. 6.1.

$$= (6356 + 300)^2 + (4800)^2$$

$$= 6730 \times 10^4$$

$$E = 8200\text{V}.$$

$$\sin\delta = \frac{IX}{E} = \frac{4800}{8200} = 0{\cdot}586$$

$$\therefore\ \underline{\delta = 35{\cdot}9^\circ \text{ electrical.}}$$

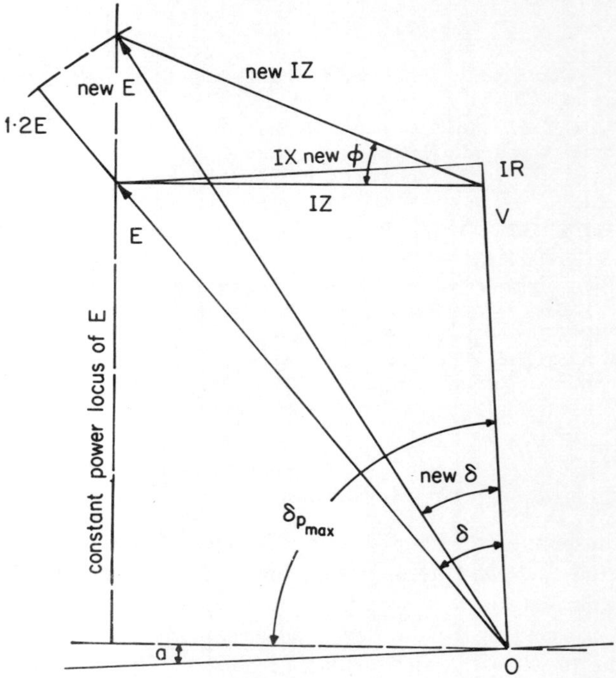

Fig. Ex. 6.1.

(b) New $E = 1{\cdot}2 \times 8200 = 9840$V.

$$\text{Now } P_o = \frac{VE}{Z}\sin(\delta + a) - \frac{V^2}{Z}\sin a$$

and $Z = (R^2 + X^2)^{\frac{1}{2}} = 8{\cdot}015 \simeq 8$, $\sin a = R/Z = 0{\cdot}0625$

$$\therefore\ P_o = \frac{6356 \times 9840}{8}\sin(\delta + a) - \frac{(6356)^2}{8} \times 0{\cdot}0625$$

$$= (7820 \times 10^3)\sin(\delta + a) - 316 \times 10^3$$

$$= VI\cos\phi = \text{constant}$$

$$= 6356 \times 600 \times 1 = 3813{\cdot}6 \times 10^3 \text{ W/phase}$$

$$\therefore\ \sin(\delta + a) = \frac{4129{\cdot}6}{7820} = 0{\cdot}53.$$

From tables, $(\delta + a) = 32°$. But $a = \sin^{-1} 0{\cdot}0625 = 3{\cdot}6°$.

$\therefore$ New $\delta = 28{\cdot}4°$ electrical.

Using the cosine rule:

$$(IZ)^2 = V^2 + E^2 - 2VE\cos\delta$$
$$= (6356)^2 + (9840)^2 - 2 \times 6356 \times 9840 \times 0{\cdot}88$$
$$= 2750 \times 10^4$$
$$IZ = (2750 \times 10^4)^{\frac{1}{2}} = 5240\text{V}.$$

$$\therefore \quad \text{New } I = \frac{5240}{8} = \underline{655\text{A}}.$$

Now Power/phase $= VI\cos\phi = 3813{\cdot}6 \times 10^3$ W (unchanged).

$$\therefore \text{New } \cos\phi = \frac{3813 \times 10^3}{6356 \times 655} = \underline{0{\cdot}92 \text{ lagging}}.$$

(c) Maximum power occurs when $(\delta + a) = 90°$.

$$\text{Then max. } P_0 = \frac{VE}{Z} - \frac{V^2}{Z}\sin a$$
$$= 7820 \times 10^3 - 316 \times 10^3$$
$$= \underline{7504 \text{ kW/phase}}.$$

Graphical: Choose a voltage scale and draw the u.p.f. phasor diagram, Fig. Ex. 6.1. Draw a line at right angles to the u.p.f. IZ. Describe an arc of radius $= 1{\cdot}2E$ to intersect the perpendicular to IZ. Join the intersection to the origin 0, which determines the position of new E. Join also to the end of the V phasor, which determines the position of the new IZ. New ϕ is obtained between old and new IZ's, and old and new δ's are obtained as shown.

IZ, I and $\cos\phi$ may be determined at the maximum power condition by swinging the new E to the position where $\delta = (90° - a)$.

EXAMPLE 6.2.

A three-phase, 10 kVA, star-connected synchronous motor operates on rated 346V a.c. mains. On load the machine takes 16·67A at unity power factor. Calculate the load angle and p.u. excitation, (a) using two-reaction theory, (b) using synchronous reactance, (non-salient pole) theory, given that $X_d = 8\Omega$, $X_q = 5\Omega$.

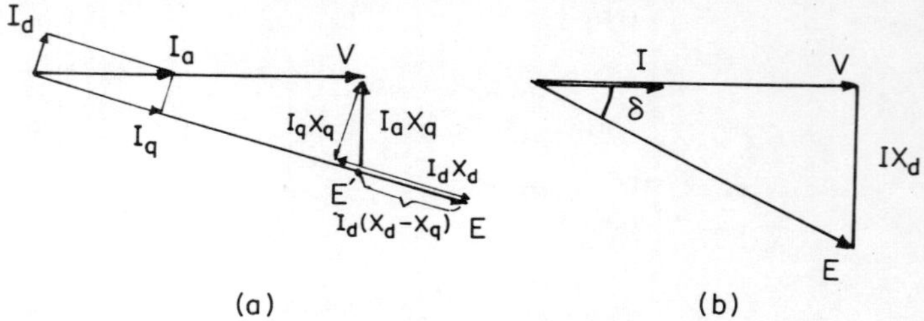

Fig. Ex. 6.2.

Solution :

With star connection : $V = 200\text{V/phase}$, $I_a = 16{\cdot}67\text{A}$.

(a) Referring to the phasor diagram, Fig. Ex. 6.2(a):

$$\begin{aligned} E' &= V - jI_a X_q \\ &= (200 + j0) - j(16{\cdot}67)(5) \\ &= 200 - j83{\cdot}35 \\ &= (40\,000 + 6930)^{\frac{1}{2}} \overline{\backslash \tan^{-1}(83{\cdot}35/200)} \\ &= 216{\cdot}2 \;\underline{/-22{\cdot}6^\circ}. \end{aligned}$$

$$\therefore \delta = 22{\cdot}6^\circ \text{ electrical.}$$

Now: $I_d = I_a \sin\delta = 16{\cdot}67 \times 0{\cdot}384 = 6{\cdot}4\text{A}$

$$\begin{aligned} E &= E' + I_d(X_d - X_q) = 216{\cdot}2 + (6{\cdot}4 \times 3) \\ &= 235{\cdot}4\text{V/phase} \end{aligned}$$

$$\text{p.u. excitation} = \frac{235{\cdot}4}{200} = 1{\cdot}177.$$

(b) Referring to phasor diagram Fig. Ex. 6.2(b)

$$\begin{aligned} E^2 &= V^2 + (IX_d)^2 \\ &= (200)^2 + (16{\cdot}67 \times 8)^2 \\ &= 4 \times 10^4 + 1{\cdot}78 \times 10^4 = 5{\cdot}78 \times 10^4 \end{aligned}$$

$$\therefore E = (5{\cdot}78)^{\frac{1}{2}} = 240\text{V/phase}$$

$$\text{p.u. excitation} = \frac{240}{200} = 1{\cdot}2$$

$$\tan\delta = IX_d/V = 133{\cdot}6/200 = 0{\cdot}668$$

$$\therefore \delta = 33{\cdot}75^\circ \text{ electrical.}$$

Note the particular discrepancy in the prediction of δ, using non-salient pole theory on a salient-pole machine.

EXAMPLE 6.3.

Establish the power/load-angle relationship for the salient-pole synchronous machine. Determine for the following six-pole salient-pole machine, the value of the power and load angle at stalling: the machine has 1 p.u. excitation and is rated at 10 kVA, 346 V, 50 c/s. It is star-connected with phase reactances $X_d = 8\Omega$ and $X_q = 4{\cdot}8\Omega$.

Solution:

First part.

The theoretical derivation is given in the text, p. 178 and it is shown, for a salient-pole machine:

$$P = \frac{VE}{X_d}\sin\delta + \frac{V^2}{2}\left(\frac{1}{X_q} - \frac{1}{X_d}\right)\sin 2\delta \ \text{W/phase.}$$

Numerical part.

Maximum power occurs when $dP/d\delta = 0$. Differentiating the above expression for power, we obtain the condition:

$$\cos\delta_m = \frac{1}{4}\left\{-\frac{EX_q}{V(X_d - X_q)} \pm \sqrt{\left[\left\{\frac{EX_q}{V(X_d - X_q)}\right\}^2 - 8\right]}\right\}.$$

Substituting the given values:

$$\cos\delta_m = \frac{1}{4}\left\{-\frac{0{\cdot}6}{0{\cdot}4} + \sqrt{\left(\left(\frac{0{\cdot}6}{0{\cdot}4}\right)^2 + 8\right)}\right\}$$

$$= 0{\cdot}425$$

$$\therefore \delta_m = 64{\cdot}8^\circ \text{ electrical.}$$

Substituting this value in the power equation gives:

$$P_m = 5000 \times 0{\cdot}904 + \frac{5000}{3} \times 0{\cdot}77$$

$$= 5809 \text{ W/phase}$$

$$= 17{\cdot}427 \text{ kW for three phases.}$$

EXAMPLE 6.4.

The per-unit reactances and time constants for a synchronous generator are as follows:

$$X_d = 1{\cdot}25 \qquad X_d' = 0{\cdot}30 \qquad X_d'' = 0{\cdot}18$$

$$T_a = 0{\cdot}25\,\text{sec} \qquad T_d' = 0{\cdot}04\,\text{sec} \qquad T_d'' = 1{\cdot}10\,\text{sec}$$

When the machine is generating rated open-circuit voltage, a symmetrical short circuit is applied to its terminals at the instant e_{phase} = zero. Derive an expression for the instantaneous phase current. Hence calculate the r.m.s. phase current 0·04 sec after incidence of the short circuit.

Solution :

Theoretical part. See text p. 191.

Numerical part.

$$\text{p.u. } X = \frac{I_R X}{V_R} \text{ where } R \text{ signifies 'rated'.}$$

$$\therefore \quad \frac{V_{pu}}{X_{pu}} = \frac{V_R}{V_R} \cdot \frac{V_R}{I_R X} = \frac{V_R}{I_R X} = \frac{I_{sc}}{I_R} = I_{sc(pu)} \,.$$

Hence p.u. values are obtained by substituting as follows:

$$I_{sc} = \frac{E_0}{X_d} = \frac{1}{X_{d(pu)}} = 0{\cdot}8 \text{ p.u.}$$

$$I' = \frac{E_0}{X_d'} = \frac{1}{X'_{d(pu)}} = 3{\cdot}33 \text{ p.u.}$$

$$I'' = \frac{E_0}{X_d''} = \frac{1}{X''_{d(pu)}} = 5{\cdot}55 \text{ p.u.}$$

$$\therefore \quad \text{p.u. } i_{ac}\,(\text{r.m.s.}) = (I'' - I')\,\epsilon^{-t/T_d''} + (I' - I_{sc})\,\epsilon^{-t/T_d'} + I_{sc}$$

$$= (5{\cdot}55 - 3{\cdot}33)\,\epsilon^{-0{\cdot}04/1{\cdot}1} + (3{\cdot}33 - 0{\cdot}8)\,\epsilon^{-0{\cdot}04/0{\cdot}04} + 0{\cdot}8$$

$$= 2{\cdot}22\,\epsilon^{-0{\cdot}0364} + 2{\cdot}53\,\epsilon^{-1{\cdot}0} + 0{\cdot}8$$

$$= 3{\cdot}885$$

$$i_{dc} = \sqrt{2}\, I''\,(\cos \omega t')\,\epsilon^{-(t-t')/T_a}$$

$$= \sqrt{2} \times 5{\cdot}55 \times 1 \times \epsilon^{-0{\cdot}04/0{\cdot}25} = 7{\cdot}85\,\epsilon^{-0{\cdot}16}$$

$$= 6{\cdot}78$$

$$\therefore \quad i_{\text{rms}} = (i_{ac}^2 + i_{dc}^2)^{\frac{1}{2}} = (15{\cdot}1 + 46{\cdot}0)^{\frac{1}{2}}$$

$$= \underline{7{\cdot}84 \text{ p.u.}}$$

EXAMPLE 6.5.

A test oscillogram of the mechanical oscillations, following a sudden application of load to a six-pole 50 c/s motor-alternator set, gave a damping ratio of 0·18 and a periodic time of 0·345 sec. The measured steady-state load angle was 16·5° elect., at a loading of 8 kW. Calculate (a) the moment of inertia of the combined rotors, (b) the velocity-dependant damping, (c) the equation of motion, (natural), (d) the expression for the instantaneous load angle.

Solution:

Given periodic time = 0·345 sec, $\therefore\ f_d = 1/0{\cdot}345 = 2{\cdot}9$ c/s.

Hence $$\omega_d = 2\pi f_d = 18{\cdot}3 \text{ rad/sec.}$$

$$\omega_n = \frac{\omega_d}{(1-\zeta^2)^{\frac{1}{2}}} = 18{\cdot}5 \text{ rad/sec.}$$

$$P_s = \text{slope of } P/\delta \text{ graph} = \frac{8000}{16{\cdot}5} = 486 \text{ W/°elect.}$$

Now, $$\omega_n = (P_s/M)^{\frac{1}{2}} \quad \therefore \quad M = P_s/\omega_n^2 = \frac{486}{(18{\cdot}5)^2}$$

$$= 1{\cdot}43 \text{ W/°elect/sec}^2 = 0{\cdot}11\, Jf/p^2$$

$$\therefore \quad \underline{J = 2{\cdot}35 \text{ kg.m}^2}$$

$$P_d = 2\zeta(MP_s)^{\frac{1}{2}} = 2 \times 0{\cdot}18 \times (1{\cdot}43 \times 486)^{\frac{1}{2}}$$

$$= \underline{9{\cdot}75 \text{ W/°elect/sec.}}$$

Hence the equation of motion can be written:

$$1{\cdot}43\, d^2\delta/dt^2 + 9{\cdot}75\, d\delta/dt + 486\,\delta = 0.$$

The expression for the instantaneous load angle is:

$$\delta = \delta_{ss}\left[1 - \frac{1}{(1-\zeta^2)^{\frac{1}{2}}}\, \epsilon^{-\zeta\omega_n t} \sin\{(1-\zeta^2)^{\frac{1}{2}}\, \omega_n t + \phi\}\right]$$

where $$\tan\phi = \frac{(1-\zeta^2)^{\frac{1}{2}}}{\zeta}.$$

Therefore, $$\delta = 16{\cdot}5\left[1 - 1{\cdot}02\,\epsilon^{-3\cdot35t} \sin(18{\cdot}3t + 79{\cdot}5°)\right]$$

N.B. An experimental check on the value of J can be made using the retardation test, (see p. 93).

7 The induction motor

The polyphase induction motor

This is the most extensively used a.c. motor in industry. The stator is wound as for an a.c. generator with a three-phase winding, distributed according to a preferred number of poles. If the stator winding is connected to a balanced three-phase a.c. supply, a rotating radial magnetic field is produced. The rotating field rotates at a synchronous speed, $n_s = f/p$ rev/sec, or $\omega_s = 2\pi f/p = \omega/p$ mech. rad/sec, where p = pairs of poles, f = supply frequency, c/s, and ω = supply frequency in radians/sec.

If slotting is neglected, the machine has a uniform airgap. The rotor may be either 'wound' or 'squirrel-cage' type and both forms of rotor winding are short-circuited during normal operation of the machine as an induction motor.

The 'wound' rotor has a winding similar to the stator winding, and electrical access to the rotor winding is by means of slip-rings and brushes. When the rotor winding is on open-circuit and stationary; (a) the machine will behave as a three-phase transformer, but with a poorer power-factor on account of the larger magnetising current due to the airgap which permits motion; (b) the phase-shift of the rotor or 'secondary' voltage relative to the supply, depends on the position of the rotor winding; connected as an auto-transformer the machine becomes an induction regulator, providing a controlled adjustable voltage from the fixed supply voltage; (c) driving the rotor provides frequency-changing facilities.

The 'squirrel-cage' rotor has uninsulated copper or aluminium rotor conductors or 'bars' embedded in the rotor core slots. The bars are short-circuited by end-rings of the same material welded to the bars. Hence the rotor winding is isolated, since no slip-ring contacts are necessary. Construction is economical and robust. The absence of live brush contacts enables the machine to be employed in otherwise unfavourable atmospheres, but restricts the control of performance by methods which modify the rotor circuit constants.

On normal operation as a motor, torque is developed due to induced currents in the short-circuited rotor winding. The rotor accelerates until stable conditions are reached.

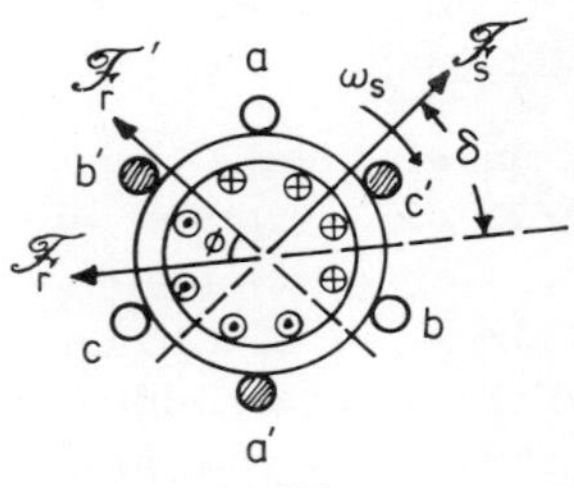

Fig. 7.1.

In the elementary case, Fig. 7.1, the instantaneous distribution of induced e.m.f's in the rotor is shown. For a pure resistance rotor circuit, the current distribution is in phase with the e.m.f. and torque is developed α $\mathcal{F}_r'\mathcal{F}_s$, since the misalignment is 90°. But the rotor circuit is reactive and the rotor current lags rotor e.m.f. by a phase angle ϕ. Hence the misalignment between $\mathcal{F}_r$ and $\mathcal{F}_s$ $= \delta = (90 - \phi)$ and, Torque $\alpha\ \mathcal{F}_r\mathcal{F}_s \sin\delta \ \alpha\ \mathcal{F}_r\mathcal{F}_s \cos\phi$

The steady-state running speed, n, is less than the synchronous (field) speed, n_s, and the difference speed, $(n_s - n)$ is termed 'slip speed'. Expressed as a fraction:

$$\frac{(n_s - n)}{n_s} = \frac{(\omega_s - \omega_r)}{\omega_s} = s$$

termed 'fractional slip'. ω_r = rotor speed, mech.rad/sec.

A steady torque depends upon a steady misalignment between rotor and stator fields, i.e. when the speed of the rotor field relative to the *stator* $= \omega_s$.

ω_s may be written $= s\omega_s + (1 - s)\omega_s = s\omega_s + \omega_r$
= (speed of rotor field relative to rotor) + (rotor mech. speed).

The frequency of induced voltages and currents in the rotor winding $= sf$ c/s, and in calculating rotor reactances, rotor radian-frequency at slip $s = s\omega$ elect. rad/sec.

Now, power in the stator rotating field = power input to rotor = rotor power output + rotor copper loss. That is:

$$\omega_s \circled{T} = \omega_r \circled{T} + \text{rotor copper loss,}$$

where Ⓣ = gross torque = nett shaft torque + friction and windage.

$$\therefore \quad \text{Rotor copper loss} = (\omega_s - \omega_r)\circled{T} = s\omega_s \circled{T}$$
$$= m\, I_r^2 R_r \quad \text{where } m = \text{phases.}$$

$$\text{Thus,} \qquad \frac{\text{rotor copper loss}}{\text{rotor input}} = \text{slip.}$$

$$\text{Now, rotor input} = \omega_s \circled{T} = \frac{mI_r^2 R_r}{s}$$

$$\therefore \quad \circled{T} = \frac{1}{s\omega_s}\, mI_r^2 R_r\,.$$

$$\text{Also, rotor output} = \frac{mI_r^2 R_r}{s} - mI_r^2 R_r$$

$$= \left(\frac{1 - s}{s}\right) mI_r^2 R_r$$

$$= \omega_r \circled{T}\,.$$

Note that, of the power delivered to the rotor across the airgap, a part $(1 - s)$ is developed as mechanical output and the remaining part (s) is lost in rotor ohmic resistance.

The equivalent circuit

Principles similar to those applied in transformer theory are adopted, the stator being viewed as the primary winding, the rotor being the secondary. Because the system is normally balanced, a PER PHASE analysis is used.

An equivalent circuit per phase of a polyphase induction motor can be drawn as in Fig. 7.2(a).

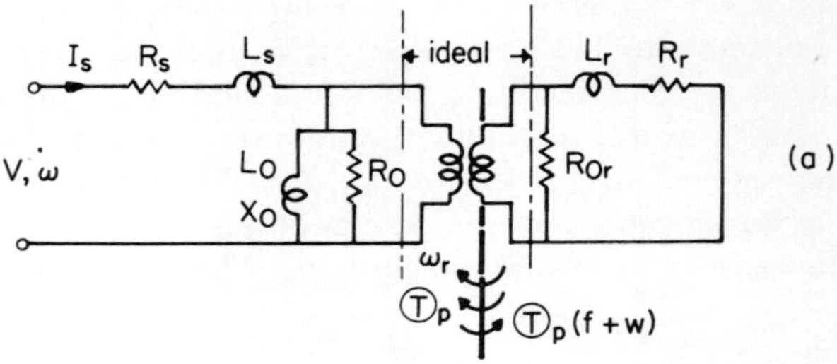

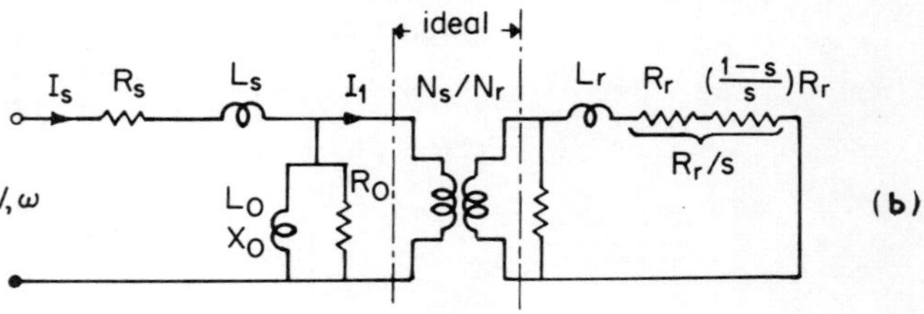

Fig. 7.2.

R_s and R_r are stator and rotor winding resistances/phase,

L_s and L_r are stator and rotor leakage inductances/phase,

where $L_s = (L_{11} - L_{12})$ and $L_r = (L_{22} - L_{12})$ in transformer notation.

L_0 = magnetising inductance = L_{12}

R_0 = resistance to represent stator core loss due to hysteresis and eddy currents.

R_{0r} = resistance to represent rotor core losses, normally negligible at low rotor frequencies.

Corresponding reactances at normal running conditions will be:

$$X_s = \omega L_s;\ X_0 = \omega L_0 \text{ and } X_r = s\omega L_r .$$

The mechanical rotor output, which is the cause of the machine running at a slip speed, can be represented as an equivalent rotor resistance element, which will dissipate an equivalent power, and as shown above:

$$\text{rotor output/phase} = I_r^2 \left(\frac{1 - s}{s}\right) R_r .$$

Hence the mechanical output is represented by a rotor resistor

$= R_r(1-s)/s$, Fig. 7.2(b). This component resistor varies with slip, whereas the winding ohmic resistance R_r does not. If the two are combined, the apparent rotor resistance $= R_r/s$.

Thus at standstill, when $s = 1$, all the induced power is dissipated as $I_r^2 R_r$ and no mechanical output occurs. When $s = 0$, the machine will operate as a synchronous machine, with d.c. (zero frequency) supplied to the rotor from an independent source.

Referred resistances and reactances

It is convenient, for purposes of test and analysis, to be able to refer the rotor resistance and reactance to the stator circuit and thereby to derive an equivalent circuit referred to the stator. This is particularly desirable for a cage rotor machine, since the rotor circuit constants are not readily accessible.

Now, referring to Fig. 7.2(b):

At standstill, $(s = 1)$,

1 volt in the rotor circuit $= N_S/N_r = k$ volts in the stator, where k = turns ratio/phase, N = turns/phase winding.

At slip s,

$$1 \text{ volt in rotor circuit} = \frac{1}{s}\left(\frac{N_S}{N_r}\right) = \frac{k}{s} \text{ volts in stator.}$$

The current transformation ratio depends on the balance of m.m.f's, and $I_r N_r = I_1 N_S$, whence $I_r = kI_1$ and is *independent of slip.*

Then $I_r X_r$ volts in the rotor circuit referred to the stator circuit, at slip s,

$$= \frac{k}{s}(I_r X_r) = \frac{k^2}{s}(I_1 X_r)$$

$$= I_1 \frac{k^2}{s}(s\omega L_r) = I_1 \omega k^2 L_r$$

$$= I_1 X'_r$$

Hence referred rotor reactance, $X'_r = \omega L'_r$ where $L'_r = k^2 L_r$.

If the same reasoning is applied to R_r we obtain a value for R_r referred to the stator circuit, at slip s,

$$= \frac{k^2}{s} R_r = \frac{R'_r}{s} \text{ where } R'_r = k^2 R_r.$$

Now at any value of slip, 1 W in the rotor = 1 W in the stator.
Then rotor *winding* copper loss $= I_r^2 R_r = k^2 I_1^2 R_r$.
For equivalence this must $= I_1^2 R_r'$.

Hence $R_r' = k^2 R_r$.

But, allowing for slip in the previous transformation it was shown that R_r referred to the stator $= R_r'/s$, and it follows that rotor power (in stator) $= I_1^2 R_r'/s$.

Thus the effect of slip is again demonstrated as an additional power

$$= I_1^2 \left(\frac{R_r'}{s} - R_r'\right) = I_1^2 R_r' \left(\frac{1 - s}{s}\right)$$

$=$ mechanical power developed by rotor.

This is an important point, since confusion can arise if the power which causes slip is forgotten.

An equivalent circuit, referred to the stator can now be drawn as shown in Fig. 7.3(a).

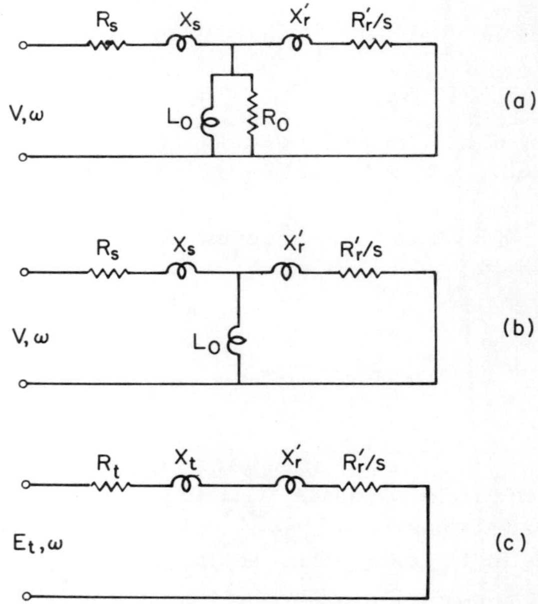

Fig. 7.3.

Simplified equivalent circuits

Various assumptions commonly adopted permit further simplification of the equivalent circuit, but at the expense of some loss of accuracy of representation and of performance prediction.

(*a*) *The Thévénin equivalent circuit.*

Assuming that R_0 may be removed and the core losses which it represents are grouped with the friction and windage losses, then the equivalent circuit can be drawn as in Fig. 7.3 (b).

The Thévénin equivalent source voltage:

$$E_t = \frac{jX_0}{R_s + j(X_s + X_0)} \text{ V}$$

and

$$R_t + jX_t = \frac{jX_0\,(R_s + jX_s)}{R_s + j(X_s + X_0)} \;.$$

Hence

$$I_1 = \frac{E_t}{(R_t + R_r'/s) + j(X_t + X_r')}$$

and the Thévénin equivalent circuit is shown in Fig. 7.3 (c).

Now, rotor power/phase:

$$P_{ph} = \frac{(I_r')^2 R_r'}{s}$$

$$= \frac{R_r'}{s}\,\frac{E_t^2}{\left[\left(R_t + \dfrac{R_r'}{s}\right)^2 + (X_t + X_r')^2\right]} \;.$$

Rotor power for m phases: $P = mP_{ph}$

and gross torque Ⓣ $= \dfrac{1}{\omega_s} P$

$$= \frac{mE_t^2 R_r'}{s\omega_s\left[\left(R_t + \dfrac{R_r'}{s}\right)^2 + (X_t + X_r')^2\right]} \text{ newton-metres.}$$

The torque/slip relation: The general shape of this relation may be determined by considering limiting tendencies, as follows. Working from the Thévénin values:

When $s \rightarrow 0$, R_r'/s becomes large and we can approximate,

$$Ⓣ \approx \frac{mE_t^2}{\omega_s R_r'} \cdot s$$

that is, Ⓣ $\propto s$ approximately.

When $s \to 1$ $\qquad R_t + \dfrac{R_r'}{s} \ll (X_t + X_r')$

and $$Ⓣ \approx \frac{mR_r' E_t^2}{s\omega_s (X_t + X_r')^2}$$

that is Ⓣ $\propto 1/s$ approximately.

Hence the curve of Ⓣ$/s$ follows a transition from Ⓣ $\propto s$ to Ⓣ $\propto 1/s$, Fig. 7.4.

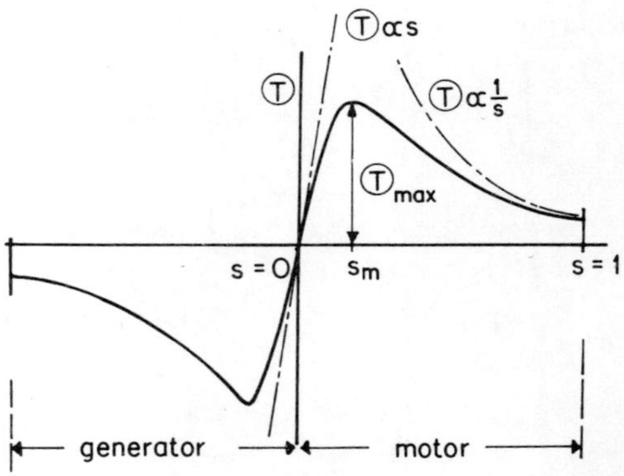

Fig. 7.4.

The value of slip at which maximum torque occurs is obtained by differentiation, and the condition is obtained when d Ⓣ$/ds = 0$. Simplifying the expression for torque we can write:

$$Ⓣ = \frac{KR_r'}{s} \frac{1}{(R_t + R_r'/s)^2 + X^2}$$

$$= \frac{KR_r'}{\left[sR_t^2 + 2R_t R_r' + \dfrac{(R_r')^2}{s} + sX^2\right]}$$

where $\qquad K = \dfrac{mE_t^2}{\omega_s}$ and $X = (X_t + X_r')$.

Then

$$\frac{d\text{Ⓣ}}{ds} = \frac{KR_r'\left[R_t^2 - \left(\frac{R_r'}{s}\right)^2 + X^2\right]}{(\text{denominator})^2} = 0 \text{ for max. Ⓣ}$$

and

$$\frac{R_r'}{s} = \pm\,[R_t^2 + (X_t + X_r')^2]^{\frac{1}{2}}$$

(Compare with maximum-power-transfer theorem, where $Z_{\text{load}} = Z_{\text{source}}$ at maximum power transfer conditions).

Hence the value of slip at which maximum torque occurs:

$$s_m = \pm\, R_r'/[R_t^2 + (X_t + X_r')^2]^{\frac{1}{2}}$$
$$= \pm\, R_r'/(R_t^2 + X^2)^{\frac{1}{2}}$$

using the simpler algebraic form.

Substituting this value in the expression for torque:

$$\therefore \quad \text{Ⓣ}_{\text{max}} = \frac{K(R_t^2 + X^2)^{\frac{1}{2}}}{\{[R_t + (R_t^2 + X^2)^{\frac{1}{2}}]^2 + X^2\}}$$

$$= \frac{K(R_t^2 + X^2)^{\frac{1}{2}}}{2(R_t^2 + X^2) + 2R_t(R_t^2 + X^2)^{\frac{1}{2}}}$$

$$= \frac{K}{2[(R_t^2 + X^2)^{\frac{1}{2}} + R_t]}\,.$$

Note that this is constant for a given supply voltage and frequency.

The possibility of the negative value of slip occurs when the rotor speed is greater than synchronous speed, i.e. generator operation. Note that Ⓣ_{max} is then greater than with positive value of slip. Fig. 7.4.

Starting torque is obtained when $s = 1$, thus:

$$\text{Ⓣ}_{\text{start}} = \frac{KR_r'}{(R_t + R_r')^2 + (X_t + X_r')^2}\,.$$

Maximum torque may be obtained at starting, when $s_m = 1$ and this condition is:

$$R_r' = [R_t^2 + (X_t + X_r')^2]^{\frac{1}{2}}$$

which is in excess of the normal R_r'. The condition is achieved by connecting extra resistance in the rotor circuit, with a wound rotor machine. For a cage-rotor machine a special starting winding has to be considered, or similar means.

The stator current. Referring to Fig. 7.3(b) this may be obtained directly :

$$I_s = \frac{V}{(R_s + jX_s) + \dfrac{jX_0\left(\dfrac{R_r'}{s} + jX_r'\right)}{\dfrac{R_r'}{s} + j(X_0 + X_r')}} .$$

(*b*) An alternative common approximation assumes the branch circuit moved to the supply terminals. Then we have:

$$I_s = I_1 + I_0$$

$$I_1 = \frac{V}{\left(R_s + \dfrac{R_r'}{s}\right) + j(X_s + X_r')} = I_r' .$$

$$\text{Output power, } P/\text{phase} = (I_r')^2\,\frac{R_r'}{s}$$

$$= \frac{V^2 \cdot \dfrac{R_r'}{s}}{\left(R_s + \dfrac{R_r'}{s}\right)^2 + (X_s + X_r')^2}$$

$$= \omega_s \, \textcircled{T}_{ph} .$$

Comparing with the Thévénin solution, maximum torque occurs at slip:

$$s_m = R_r'/[R_s^2 + (X_s + X_r')^2]^{\frac{1}{2}} .$$

The solutions will be seen to be the same except that V replaces E_t and R_s and X_s replace R_t and X_t. The approximation depends upon R_0 and X_0 being $\gg R_s$ and X_s. Discussion of the torque/slip characteristics is as given for the Thévénin case, with the stated algebraic changes.

This approximation leads to the geometrical prediction method known as the Circle Diagram which has become less popular with the advent of computers. Further considerable simplification occurs if R_s and X_s are considered negligible, but at the expense of loss of accuracy of prediction.

Measurement of parameters

The induction motor offers a popular means of illustrating difficulties inherent in attempting to obtain an 'exact' equivalent circuit of a machine. Simple tests which employ terminal measurements are preferred.

Consider the equivalent circuit, Fig. 7.2(b). The problem is to obtain representative values for the various 'constants' indicated. The stator winding ohmic resistance, R_s, can be obtained by a d.c. volt-drop test. For a wound-rotor machine R_r may similarly be determined and the turns ratio, $N_s/N_r = k$, is obtained by open-circuit volt-ratio tests. The voltage ratio is obtained for stator-rotor and for rotor-stator giving mean $k = [(V_s'/V_r)(V_s/V_r')]^{\frac{1}{2}}$ where the primed quantity indicates the input voltage.

The following tests may be considered:

(a) If a suitable driving machine is available, the induction motor under test is driven at synchronous speed. Then $s = 0$, and $R_r'/s = \infty$, i.e. open-circuit, and the equivalent circuit is reduced in complexity, Fig. 7.3(a). Readings of applied stator voltage, current and power enable the equivalent impedance, $Z = R + jX$, as 'seen' by the supply to be determined. Referring to Fig. 7.3(a), for the shunt branch:

$$\frac{jR_0X_0}{(R_0 + jX_0)} = \frac{jR_0X_0(R_0 - jX_0)}{(R_0^2 + X_0^2)}$$

$$= \frac{jR_0^2X_0 + R_0X_0^2}{(R_0^2 + X_0^2)}.$$

Hence
$$R = R_s + \frac{R_0X_0^2}{(R_0^2 + X_0^2)}$$

$$X = X_s + \frac{X_0R_0^2}{(R_0^2 + X_0^2)}.$$

A similar set of equations involving R_r and X_r can be obtained if the induction machine is driven at synchronous speed, but is rotor-fed and the stator is short-circuited. This possibility is naturally restricted to wound-rotor machines.

In either case it is necessary to obtain further information to separate X_s and X_0. X_s may be determined by calculation from design detail, or by using a search coil on the machine. Alternatively a Standstill Test, (see below) may be carried out. With a cage-rotor machine a standstill test would be necessary to determine the rotor resistance and reactance, referred to the stator.

(*b*) The no-load or light-running test. If a separate driving machine is not available, the induction motor is run light, (without a shaft load), at normal applied voltage and frequency. Readings of input voltage, current and power per phase enable the no-load impedance, $Z_n = (R_n + jX_n)$, to be determined.

Since the speed is very nearly synchronous, we derive as in (a) above:

$$X_n = X_s + \frac{X_0 R_0^2}{(R_0^2 + X_0^2)} \,.$$

Now $R_0 \gg X_0$ and we can assume:

$$X_n = (X_s + X_0)\,.$$

The standstill test described below enables these parameters to be separated.

The difference between the power input on light-running and the ohmic loss in the stator winding resistance can be taken as the core, friction and windage loss. Hence, if this test is carried out for a range of reduced applied voltages, extrapolation of the power/voltage graph to zero voltage enables an estimate of the friction and windage, (mechanical), loss to be made. (cf. p. 93). The small variation in speed is considered negligible.

(*c*) The standstill test. A reduced stator voltage is applied and the short-circuited rotor is moved slowly to a position of mean current, where it is clamped. This is taken as the position of mean gap reluctance.

As in the case of a transformer *s*/*c* test, at reduced voltage the core loss watts may be considered negligible. Hence $W_{sc} = 3I_{sc}^2 R_e$ where $R_e = R_s + R_r'$. A volt drop test enables R_s to be determined so that R_r' can be obtained.

Since $V_{sc}/I_{sc} = Z_e$, the effective reactance X_e is determined and this is assumed $= (X_s + X_r')$, because at the reduced voltage the magnetising branch circuit is assumed negligible. Generally it is assumed $X_s = X_r'$.

These results correspond to supply frequencies and are appropriate for conditions at start. Effective values for running conditions may be estimated by reduced frequency tests, for example the A.I.E.E. Test Code suggests tests at 15 c/s for 60 c/s motors. Otherwise values may be computed by assuming $X \propto f$. For small motors the effects of frequency are negligible.

Elementary analysis

An elementary analysis is obtained if stator resistance and reactance volt drops are neglected. Emphasis is then placed on the role of rotor impedance in relation to machine performance. The approximation may be acceptable for large machines, but for small machines errors of prediction up to 10 per cent may be incurred. Using this approximation:

At standstill, $V = kV_r$; at slip s, $V_r = \dfrac{sV}{k}$.

Also, remember $X_r = s\omega L_r$

$$\therefore \quad I_r = \frac{V_r}{(R_r^2 + X_r^2)^{\frac{1}{2}}} = \frac{sV}{k(R_r^2 + X_r^2)^{\frac{1}{2}}}.$$

$$\text{Now,} \quad \text{rotor intake} = \frac{\text{rotor copper loss}}{\text{slip}}$$

$$\text{or} \quad \omega_s \, Ⓣ = \frac{mI_r^2 R_r}{s}.$$

Substituting for I_r and rearranging:

$$Ⓣ = \frac{m}{\omega_s} \cdot \frac{s^2 V^2}{k^2 (R_r^2 + X_r^2)} \cdot \frac{R_r}{s}$$

$$= \frac{m}{\omega_s k^2} \cdot \frac{sV^2 R_r}{(R_r^2 + X_r^2)} \quad \text{newton.metres.}$$

This expression is commonly simplified to:

$$Ⓣ \propto \frac{sV^2 R_r}{(R_r^2 + X_r^2)} = \frac{s^2 V^2 R_r}{R_r^2 + (s\omega L_r)^2}.$$

To obtain the condition for maximum torque:

$$\frac{\mathrm{d}Ⓣ}{\mathrm{d}s} = \frac{[R_r^2 + (s\omega L_r)^2]\, V^2 R_r - sV^2 R_r (2s\omega^2 L_r^2)}{(\text{denominator})^2} = 0 \text{ at } Ⓣ_{\max}.$$

This gives the condition, $R_r^2 = s^2\omega^2 L_r^2$

or $R_r = \pm\, s\omega L_r$.

Then, slip at maximum torque, $s_m = \dfrac{R_r}{\omega L_r} = \dfrac{\text{rotor resistance}}{\text{standstill rotor reactance}}$

Also, $$Ⓣ_{max} = \frac{m}{\omega_s k^2} \cdot \frac{sV^2(s\omega L_r)}{2(s\omega L_r)^2} = \frac{mV^2}{2k^2\omega_s \omega L_r}$$

$$= \text{constant for constant } V \text{ and } f.$$

The Circle Diagram

Several cases arise where a machine, as a 'black box', may be broadly approximated to a reactance in series with a variable resistance and the resistance represents the energy-dissipating load.

Consider a simple series RL circuit where L is assumed constant and R is variable. Impedance $Z = (R^2 + X^2)^{\frac{1}{2}}$ and is variable.

Now, $$I = \frac{V}{Z} = \frac{V}{X} \cdot \frac{X}{Z} = \frac{V}{X}\sin\phi$$

where $\tan\phi = X/R$.

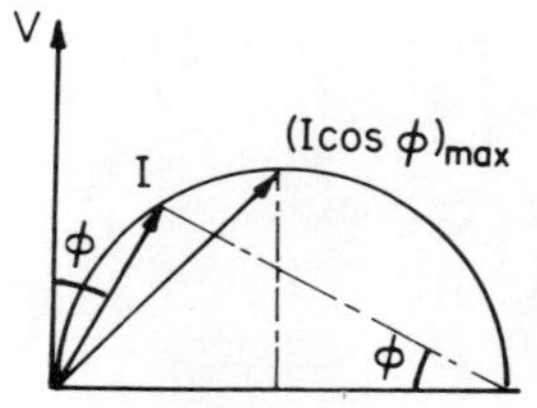

Fig. 7.5.

When V and X are constants, the current phasor will follow a circular locus, Fig. 7.5. Note that vertical distances on the diagram represent in-phase current, ($I\cos\phi$), and for constant V, vertical distances $\propto$ power, ($VI\cos\phi$). Maximum power occurs when $\phi = 45°$ or $\cos\phi = 0{\cdot}707$ lag.

The practical Circle Diagram

For an induction motor we note that at 'no load', (light running), current and losses occur and that losses occur in rotor and stator resistances on load. Hence the above idealised proposal must be modified.

Results obtained of watts/phase and current/phase from (a) A No-load, (Light running) test at rated voltage, and (b) A Short-circuit, (Standstill) test at reduced voltage, are considered; working PER PHASE, (Fig. 7.6.):

No-load current $=$ I_0 at phase angle ϕ_0 lagging.

Now $I_0 Q$ $=$ energy component of I_0 α core and friction losses, (assumed const.).

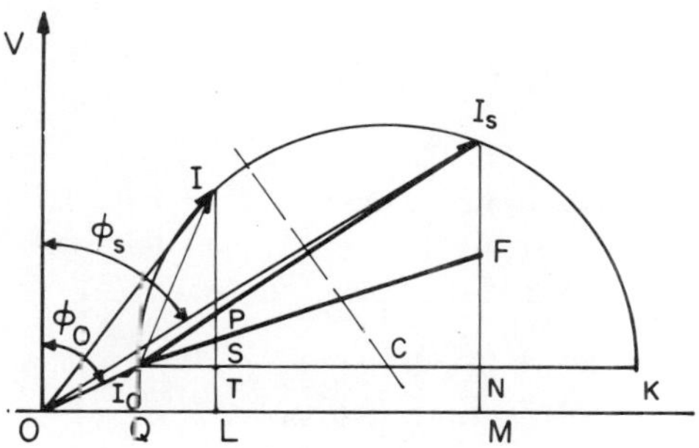

Fig. 7.6.

The standstill test is carried out at several reduced voltages, to justify the determination of I_s at normal rated voltage by direct proportion. ϕ_s is assumed constant. Hence I_s at ϕ_s lagging can be drawn.

I_0 and I_s are assumed to lie on a circular locus. The centre, C, is obtained at the intersection on $I_0 K$ of the $\perp$ bisector of the chord $I_0 I_s$, and a semi-circle is drawn.

At standstill, $I_s M$ α total power dissipated

$$NM = I_0 Q = \text{const. losses}$$

$\therefore$ $I_s N$ α total copper losses, part rotor part stator.

Divide $I_s N$ at F, where $I_s F/FN = \dfrac{\text{rotor cu. loss}}{\text{stator cu. loss}}$. Join FI_0.

We now examine the diagram for a general case of load output.

General case:

Let OI represent the current for some arbitrary load.

Draw $IL \perp OM$, and mark the intersections P, S, T.

All quantities are referred to the stator or input. Thus:

$$\text{Rotor current} = I_0 I$$

$$\text{Rotor cu. loss} = (I_0 I)^2 R_r$$

$$= (I_0 T)(I_0 K) R_r \text{ by Pythagoras theorem.}$$

Then, by similar triangles,

$$= \frac{(I_0 N)(PS)(I_0 K) R_r}{(I_s F)}$$

$$= \frac{(I_0 I_s)^2 (PS) R_r}{(I_s F)}, \text{ by Pythagoras.}$$

$$\therefore \quad \frac{(I_0 I)^2 R_r}{(PS)} = \frac{(I_0 I_s)^2 R_r}{(I_s F)}.$$

Thus PS represents rotor copper loss at load OI. Similarly, it follows that ST represents stator copper loss for the general case. Then, with appropriate power scale:

$$\text{power input/phase} = IL,$$

$$\text{rotor cu. loss} = PS,$$

$$\text{stator cu. loss} = ST,$$

$$\text{constant loss} = TL.$$

Hence the distance IP = output, and the line $I_0 I_s$ is termed the OUTPUT line. Further, Efficiency = IP/IL and Slip = PS/SI.

Now, rotor input = output + R_r cu. loss = SI = ω_s (torque)

$\therefore$ $SI \propto$ Torque and $I_0 F$ is termed the TORQUE line.

Scales: Let 1 cm = n amps.

Then, vertically, 1 cm = Vn watts, where V = phase volts.

1 cm = Vn/ω_s newton-metres.

Note also the maximum output and maximum torque are determined at the points where the tangents to the circle, drawn parallel to the output and torque lines respectively, meet the circle.

The circle diagram is a popular and convenient method of obtaining an approximate prediction of machine performance. As in the case of the simplified equivalent circuit, errors in prediction are greater for small machines.

Starting methods

(*a*) *Direct-on-line switching*

A condition of high current at low power factor exists until the machine has accelerated and built up a 'back voltage'. Temperature rise is rapid and the motor can be damaged if the start is delayed due to starting up under load. The torque at starting (*s* = 1), is inherently small, Fig. 7.4.

With large machines the heavy supply current can cause undesirable voltage regulation of the supply, causing a further deterioration of starting torque, (Ⓣ $\propto$ V^2).

The method is usually only employed with small machines, (< 5h.p.).

(*b*) *Reduced voltage methods*

(*i*) *The star/delta start.* This is a popular and economical method. The machine stator winding is arranged in star-connection at start and switched to delta-connection for running, Fig. 7.7. The reduction of supply voltage at start = $1/\sqrt{3}$ and hence starting torque = 1/3 (torque on direct switch).

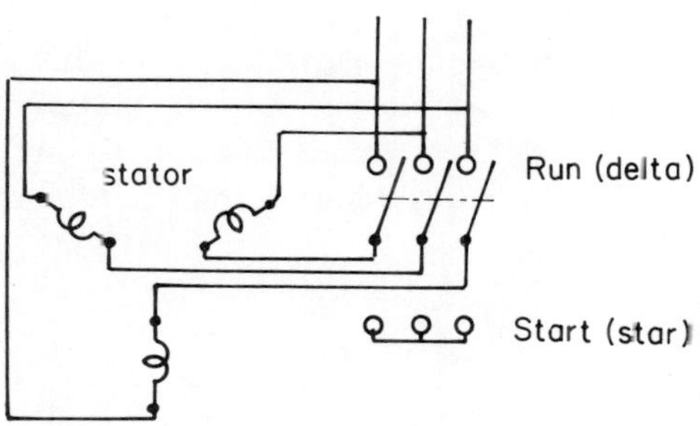

Fig. 7.7.

(*ii*) *Auto-transformer start.* The principle is the same as in (*i*) but control of the ratio of reduction, (1/*k*) is possible. Then :

$$Ⓣ_{\text{start}} = \left(\frac{1}{k}\right)^2 Ⓣ_{\text{direct-sw}} .$$

If the limiting factor is current from the supply there is advantage that voltage step-down to the machine is current step-down to the supply. This consideration also applies to the Star/Delta method.

(*c*) *Resistance starters*

(*i*) A stator resistance starter gives improved power factor, but has the disadvantages of reduced torque due to reduced voltage. It is of heavy-current rating and the benefit of current transformation which occurs in methods (b) (*i*) and (*ii*) is not obtained.

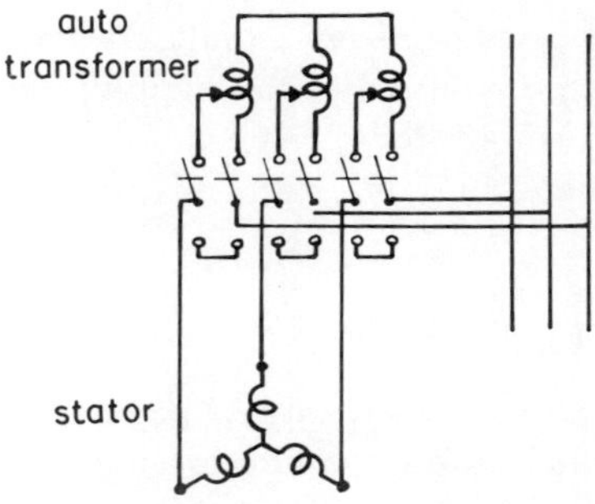

Fig. 7.8.

(*ii*) Rotor starting resistance. This method applies essentially to the wound-rotor machines, where access to the rotor impedance is possible. If the rotor resistance is increased initially so that the effective rotor resistance equals the standstill rotor reactance, (p. 215) the condition of maximum torque at starting is obtained, (Fig. 7.9). The external resistance is gradually cut out as the machine accelerates, thereby maintaining optimum torque conditions.

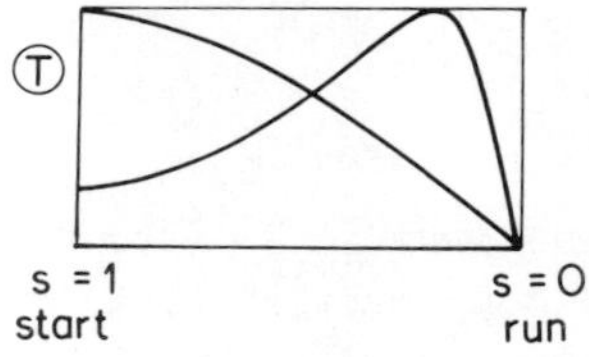

Fig. 7.9.

The principle of changed rotor resistance is obtained in cage rotor machines by using deep rotor bars. The inner part, being embedded in iron, has high reactance at the slip frequency at start. In addition eddy current effects mean that the resistance at start is high. This decreases as slip decreases, providing better torque conditions over the starting period. Double-cage rotors are a further example of these principles. But they are more expensive and have lower efficiency, although they give rapid starting acceleration.

Effects due to slotting

Two effects which are of importance with cage rotor machines arise from the effect of slotting on both sides of the airgap.
(a) If the number of rotor and stator slots is the same or has a common factor, the teeth may tend to line up in the position of minimum reluctance. This phenomenon is termed 'magnetic locking'.
(b) Because the stator winding is distributed, the stator flux can contain pronounced odd-harmonic components, notably 3rd, 5th and 7th. The 3rd harmonic components due to the three phases cancel, but the 5th harmonic wave rotates backwards, the 7th rotates forwards. Hence there are two parasitic torques superimposed on the torque/speed graph, as shown in Fig. 7.10. If the frictional torque of the motor is located in the region shown, the motor will 'crawl' at this speed under stable conditions. With a pronounced 7th harmonic the machine rotor resistance should be increased to enable a torque to be developed which is greater than the friction torque at this speed, $\omega_s/7$. The problem does not arise with wound-rotor machines having rotor-resistance starters, since the starting torque obtained is much greater than the frictional torque.

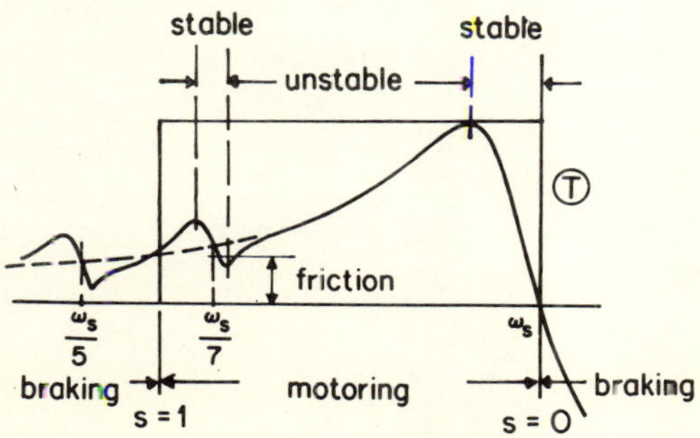

Fig. 7.10.

Speed control

The speed of an induction motor $= (1 - s)n_s$, where n_s = synchronous speed $= f/p$. Since s is normally very small, the inherent running speed is nearly constant and close to synchronous speed. Because the machine type is cheap and robust, much thought and ingenuity have been applied to the problem of obtaining variable speed characteristics. The inherent dependance on supply frequency suggests control of speed, at the stator, by (*i*) change of supply frequency, (*ii*) change of pole pairs. Control of speed at the rotor implies limitation to wound-rotor machines, where access to the rotor is possible.

(*a*) *Rotor methods*

(*i*) *Resistance control*. The torque/slip relation can be expressed in approximate form:

$$Ⓣ \propto \frac{sV^2R_r}{R_r^2 + (s\omega L_r)^2}\,.$$

At running speed, $s\omega$ is small and Ⓣ $\propto s/R_r$ approx., or at a given Ⓣ, $s \propto R_r$. Hence control of speed can be obtained with a rotor controller, approximately proportional to resistance. The rheostat is continuously rated, and the I^2R loss means increased losses and lower efficiency. Liquid (water) resistance controllers are used with large machines.

(*ii*) *Rotor injection principles.* Basically this method involves injecting a voltage of suitable frequency and phase, thereby providing the excitation and 'back' voltage from an external source. The principle and its development are described below, p. 230.

(*iii*) *The cascade connection.* With a wound-rotor machine, the slip power may be fed to a second machine. If the rotors are mechanically coupled:

$$n_1 = f/p_1 = \text{synchronous speed of 1.}$$

If n = speed of set, then slip speed $= (n_1 - n)$ and rotor 1 frequency $= (n_1 - n)p_1$.

Hence syn. speed $$n_2 = \frac{(n_1 - n)p_1}{p_2}$$

$$\therefore \quad n \approx n_2 = \frac{(n_1 - n)p_1}{p_2}$$

$$= n_1 p_1 / (p_1 + p_2).$$

This presupposes that the torques are in the same direction.

Thus four speeds are possible:

Cumulative cascade, $n = f/(p_1 + p_2)$,
Differential cascade, $n = f/(p_1 - p_2)$, providing $p_2 \neq p_1$,
Machine 1 alone, $n = f/p_1$,
Machine 2 alone, $n = f/p_2$.

Variations are possible by coupling the rotors by belt or gears, and by combining the method with pole-change arrangements, (see below).

(*b*) *Stator methods. Frequency changing.*

(*i*) Methods employing variable frequency supplies derived from rotating machines are generally limited to occasions where the a.c. generator supplies only the induction motor concerned. Multiplicity of machines adds to the first cost and to problems of maintenance, as well as being restricted in the range of frequency available.

(*ii*) *Variable-frequency thyristor invertors.* For speed control over a wide range, a more practical system involves a rectifier followed by a variable-frequency invertor. Initially the thyratron, and now superseded by a silicon thyristor with triggering facilities, has enabled a controllable frequency supply to be obtained. (cf. p. 115)

The principle of switching the thyristor is termed 'commutation', whereby it ceases to conduct. This can be natural commutation where it ceases to conduct due to a natural decrease of current to zero. Forced commutation is obtained by connecting a charged capacitor in parallel with the thyristor, thereby applying a reverse voltage across it, Fig. 7.11(a). An auxiliary d.c. supply may be provided to combat dependance of the capacitor charge upon load current.

The sequence of the firing pulses can be controlled by a pulse generator in any desired manner, and the thyristors are generally fired in diagonal pairs, resulting in a stepped waveform,

Figs. 7.11(b), 7.12. The speed-defining frequency is the fundamental of this waveform.

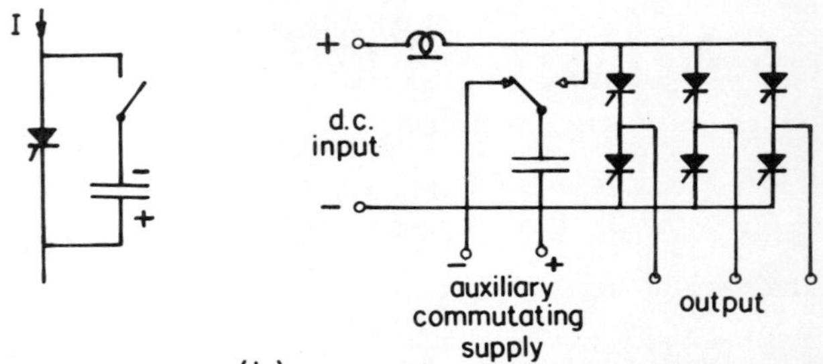

(a) (b) basic 3-phase bridge invertor circuit

Fig. 7.11.

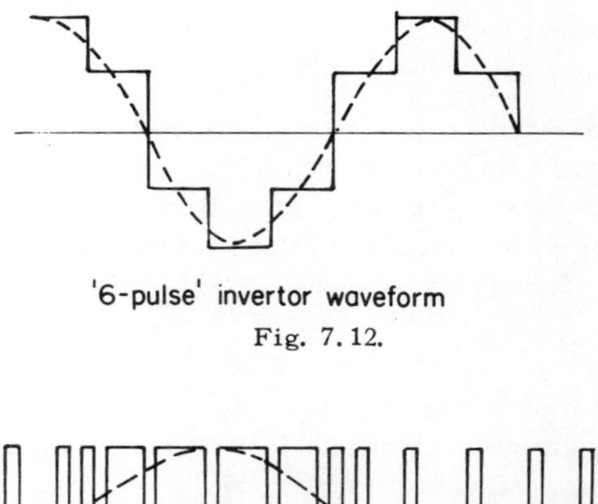

'6-pulse' invertor waveform

Fig. 7.12.

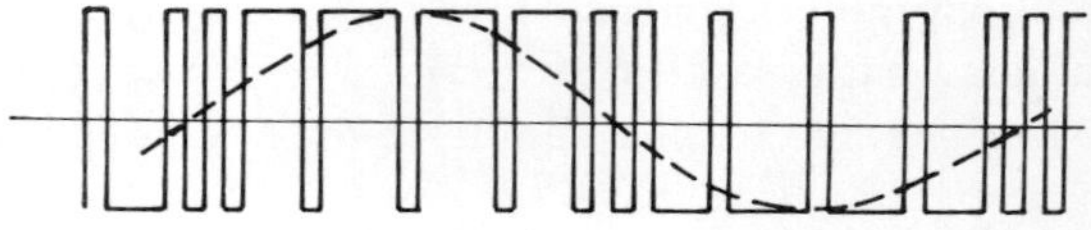

'h.f. chopping' invertor waveform

Fig. 7.13.

An alternative type of output is obtained by 'high frequency chopping', resulting in a stepped waveform as shown in Fig. 7.13. Again the fundamental frequency is the important one.

Loss of efficiency due to non-sinusoidal waveform appears to be greater than that due to the Invertor. But the method has a higher efficiency than rotating machine methods, and provides a much wider range of frequency control, (Ref. 6). At very low frequencies some pulsation of torque may occur. At higher frequencies, (400 c/s), there is promise of reduction in the size of the motor.

(*c*) *Stator methods. Pole changing.*

(*i*) Pole change in the ratio 2/1. This can be obtained for a given machine by using the 'consequent pole' principle, wherein the polarity of one group of the coils is reversed by switching relative to the other group. The elementary principle is indicated in the simple two-pole, one-phase case, Fig. 7.14.

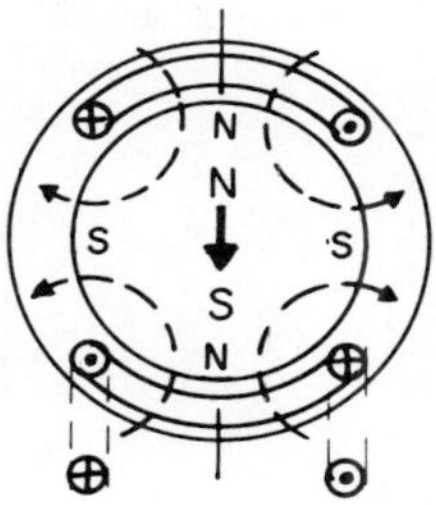

'consequent - poles' principle

Fig. 7.14.

Increased flux leakage with the larger number of poles results in a poorer power factor. The method is best suited to machines with large diameter to minimise this leakage.

(*ii*) Two separate windings may be provided. These may be located in the same slots but distributed for different pole-pairs. The design is more costly and involves deeper slots. A combination of methods (*i*) and (*ii*) can provide four different synchronous speeds, e.g. at 50 c/s,

winding 1, –	4 poles	–	1500 r.p.m.
	8 poles	–	750 r.p.m.
winding 2, –	6 poles	–	1000 r.p.m.
	12 poles	–	500 r.p.m.

In traction applications, intermediate speed control can be made by resistance controller. In Europe some 33 c/s and $16\frac{2}{3}$ c/s supplies have been used in traction systems to aid this problem of speed control. Important new methods using 'pole-amplitude modulation' and 'phase-mixing techniques' should be noted, (Refs. 1,2).

Rotor injection principles and variable speed a.c. motors

Inherent disadvantages of the polyphase induction motor are poor power factor and difficulty of uniform speed control. The development of a variable speed a.c. motor which does not possess these limitations is important, particularly a design which retains the robustness and simplicity of manufacture of the induction motor. One way of obtaining improved power factor and uniform speed control is based on the principle of rotor injection.

Power factor and speed:

Consider the simple phasor diagram of the induction motor, Fig. 7.15. It is useful here to keep in mind the transformer nature of the machine.

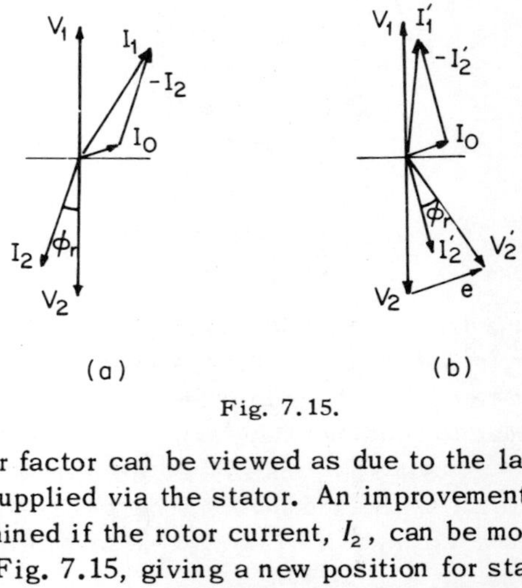

Fig. 7.15.

Poor power factor can be viewed as due to the large magnetisin current, I_0, supplied via the stator. An improvement of power factor is obtained if the rotor current, I_2, can be moved to a position I_2', Fig. 7.15, giving a new position for stator supply current, I_1', being the resultant of I_0 and the referred I_2'. Now if ϕ_r, the impedance angle of the rotor, be assumed constant it is implied that V_2 should occupy a new position, V_2'. The desired effect is obtained if a voltage e is injected into the rotor, this voltage being virtually in quadrature with the rotor current I_2'.

Speed is related to the required magnitude of rotor induced voltage, V_2, which in fact is dependant on the slip speed. Under stable running conditions V_2 circulates I_2, necessary to develop the required torque. With a rotor voltage supplied by injection from an external source, the machine runs more slowly, or faster depending on whether the injected voltage opposes or assists the purpose of V_2.

In both cases the injected voltage must be of the correct slip frequency for addition. The requirements may be summarised:

(*i*) for phase advance, an injected voltage, *e*, of slip frequency, in quadrature with I_{rotor},

(*ii*) for speed control, an injected voltage, *e*, of slip frequency, in phase aiding or opposing normal induced rotor voltage V_2.

Note that, supplied via the rotor, the necessary volt-amps for phase advance will be much less, being at slip frequency, than the volt-amps necessary at mains frequency, i.e. if supplied via the stator.

The synchronous motor is a particular case of rotor injection where the rotor, (field), is entirely supplied from an external source of d.c., i.e. at the appropriate zero slip frequency. Variation of rotor d.c. volt-amps results in a much greater variation of stator volt-amps. In this sense the machine is an amplifier.

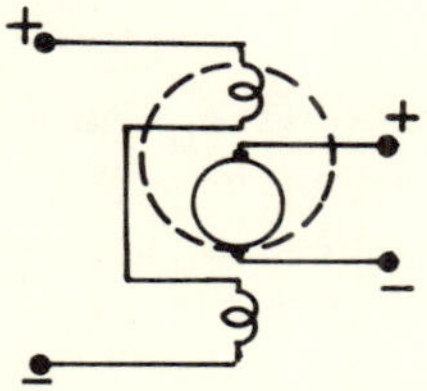

Fig. 7.16.

Phase advancers for rotor injection

Consider first a d.c. armature with polarised two-pole field, Fig. 7.16. The voltage distribution, fixed in space, about the commutator may be drawn as a graph with the negative brush as reference, and will take the form of a lobe. If a.c. is supplied to the field coils the brush voltage will alternate at the same frequency. In general, the magnitude of the generated voltage will still depend on the speed of rotation of the armature.

M. Leblanc (1895), devised a scheme for phase advance based on this principle and probably best appreciated as shown below in a two-phase case, Fig. 7.17. Two similar d.c. generators, 1 and 2, supply the corresponding phases of a two-phase induction motor rotor, but each generator has its field coil excited by the rotor current of the other phase. Each advancer has a generated voltage of slip frequency, but in quadrature with the current of the phase it feeds.

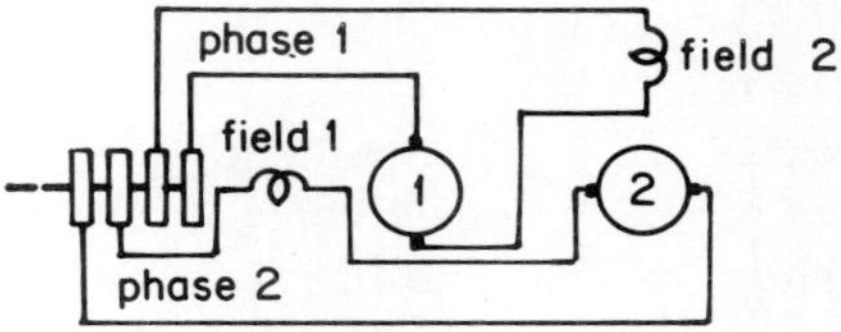

Fig. 7.17.

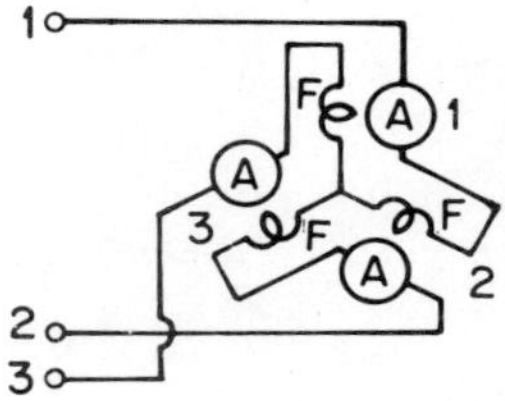

Fig. 7.18.

For the three-phase case, three small d.c. generators are required, each excited from the next phase, Fig. 7.18. In practical form the three small machines can be resolved into one machine. Referring to the principle shown in Fig. 7.16, if the d.c. armature is excited by a three-phase rotating field, the voltage distribution will rotate at the field speed, (like a hoop on a pencil). Three brushes spaced at 120° about the commutator will give a three-phase voltage output of the field frequency and they can be connected in series with the appropriate field coils to the a.c. supply, in the case of a phase advancer to the slip-rings of the induction motor, Fig. 7.19. Note that the physical position of the field coils relative to the brushes determines the phase of the brush voltage with respect to the exciting current, (see Walker below).

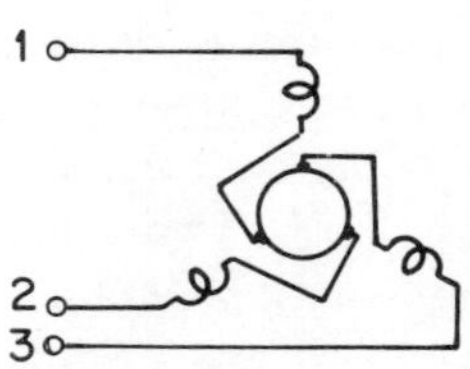

Fig. 7.19.

Contribution by Scherbius

A further development is possible since the external field coils may be dispensed with. The three sections of the armature winding constitute a delta-connected winding. This delta will create a synchronous field with respect to the armature when the brushes are supplied three-phase from, say the induction motor slip-rings. When the armature is rotated, a 'back' voltage, E_b, is generated in the armature winding. At speeds less than synchronous E_b lags I_r by 90°; at speeds greater than synchronous, E_b leads I_r by 90° since the relative direction of cutting flux is reversed, (remember induced voltage and its flux are in quadrature).

The remaining stator iron, when the field coils are removed provides only a low reluctance flux path, and could be rotating. It can be dispensed with entirely if the armature winding is located in deep closed slots, Fig. 7.20.

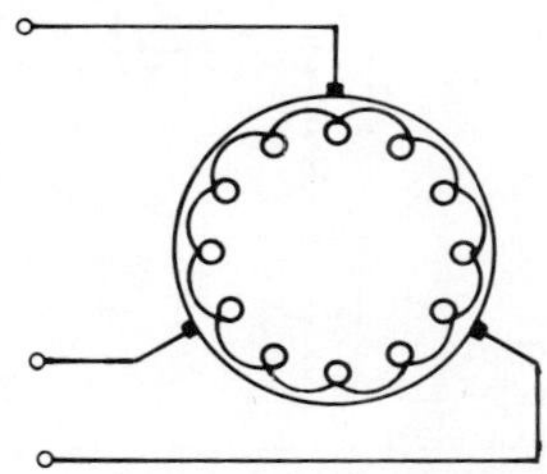

Fig. 7.20.

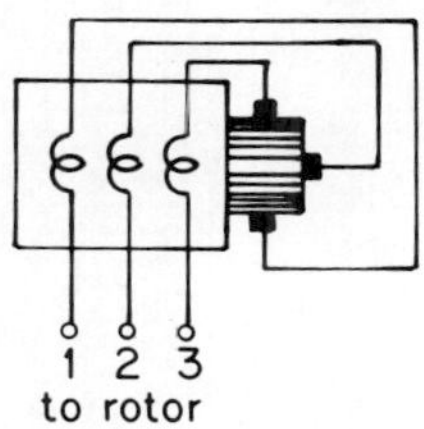

Fig. 7.21.

Contribution by Miles Walker

Re-introduction of the stator winding in series, Fig. 7.21 was proposed since it provides a valuable means of adjustment of the magnetic axis, and hence of the phase of the advancer voltage with respect to the exciter current, I_r. The disadvantage of this arrangement is that, on light loads when the power factor is poorest, rotor current is small and so the flux and injected voltage are small.

Further contribution by Scherbius

The disadvantage of Walker's arrangement can partly be reduced if the stator winding is connected in shunt, Fig. 7.22. This arrangement gives constant flux since, flux is directly proportional to rotor slip voltage, and inversely proportional to slip frequency. A reduction in load reduces slip frequency but also reduces slip voltage, giving constant flux.

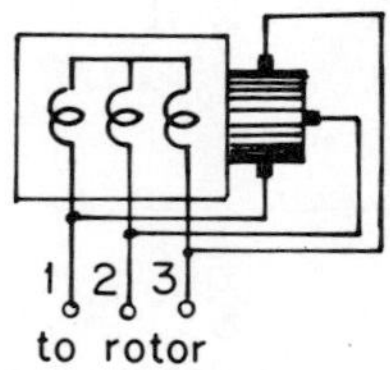

Fig. 7.22.

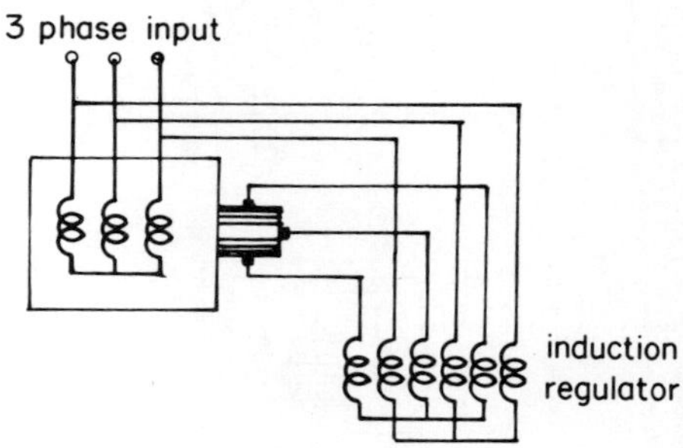

Fig. 7.23.

The three-phase commutator motor

Following the above discussion, the principle of the three-phase commutator motor is implied in Fig. 7.23. The three lines may be connected to mains supply. By varying the voltage applied to the commutator the speed can be controlled, (i.e. the armature 'back voltage' is being supplied from another source). This may be done by induction regulator, as shown by B. Schwarz, Ref. 5.

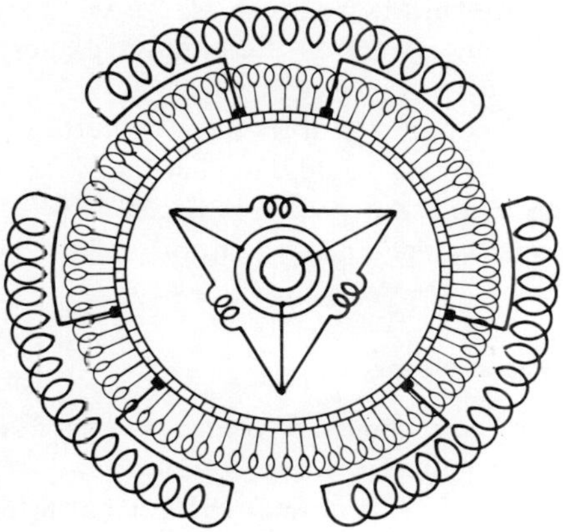

Fig. 7.24.

The Schrage motor

This is a variable-speed a.c. motor which embodies the principles of rotor injection for an induction motor. The rotor carries a three-phase winding fed via slip-rings, and also a d.c. type winding with commutator. The stator three-phase windings are fed from pairs of adjustable brushes on the commutator, Fig. 7.24.

In general the machine operates as an 'inverted' induction motor, i.e. the rotor is supplied whilst the stator is short-circuited. The primary or rotor three-phase winding provides a rotating field, being supplied from three-phase mains. When the brush pairs are closed, the stator is on short-circuit and the machine runs as an induction motor. Opening the brush pairs varies the magnitude of the injected voltage and controls speed. Rocking the brush assembly as a whole alters the phase of the injected voltage and, in general, controls power factor. In practice the brush assembly is set at an optimum position for power factor and facility for brush spread only is provided, i.e. variable speed control.

Induction machines for special purposes. (Ref. 7).

Preoccupation with the conventional cylindrical forms of induction motors is largely due to the advantages of high efficiency and the economy and simplicity of construction which arise from the symmetrical form used. When variable speed is desired, it has been seen generally to involve additional equipment or expensive changes in design. This is because of the inherent dependance of speed in induction motors on the supply frequency and the number of pole pairs.

Unconventional departures from the symmetrical form have been explored in attempts to devise, basically, a variable-speed design. Efficiency in the conventional sense is sacrificed, but this may be compensated for by the nature of the performance achieved.

The linear induction motor

Several interesting and important features, associated with asymmetry, become apparent when development of a linear induction machine is considered. If a normal cylindrical rotor machine is cut at a point on the periphery and laid flat, the phenomenon of the 'travelling field' is retained. However, (a) torque production ceases when the 'rotor' moves a distance equal to its original airgap periphery. Hence an extended stator or rotor winding must be envisaged, to provide continuing linear travel. These considerations give rise to the possibilities of (*i*) a short-rotor, (*ii*) short-stator linear machine types. (b) Unbalanced magnetic pull, conveniently neutralised in the conventional machine, provides a magnetic pull $\propto B^2$, which is much in excess of the linear 'motion producing' thrust ($= Bli$). This problem may be countered by using a double-sided rotor, (or stator), and maintaining equal airgaps on each side. (c) Edge transient effects occur due to the entry and exit of leading and trailing edge teeth with respect to the travelling field. Establishment of flux in a tooth cannot occur without an induced current in the *s/c* turn about the tooth. The result is a 'rotor' current and rotor I^2R loss even at synchronous speed.

The linear motor is not a recent invention, but it has received renewed attention as one of a number of unconventional machine types. The principle has been used in electric traction, in conveyor-belt drives, for high-speed shuttle propulsion in the weaving industry and in liquid-metal pumps.

A machine may be judged in terms of a 'Goodness Factor', Ref. 7, which is derived as follows:

If resistance of the electric circuit and reluctance of the magnetic circuit be taken as indications of 'bad' qualities the adverse qualities may be stated as proportional to (resistance × reluctance). Goodness is given by the reciprocal, viz:

$$G \propto \frac{1}{R} \times \frac{1}{\Re} \propto \left(\frac{a}{\rho l} \times \frac{\mu A}{L}\right).$$

Dimensional analysis gives: $G \propto \dfrac{1}{[\mu L T^{-1}]} \times [\mu L] \propto [T]$.

By introducing frequency, (or speed), the factor becomes a numeric:

$$G = \frac{\mu A a}{\rho L l} f.$$

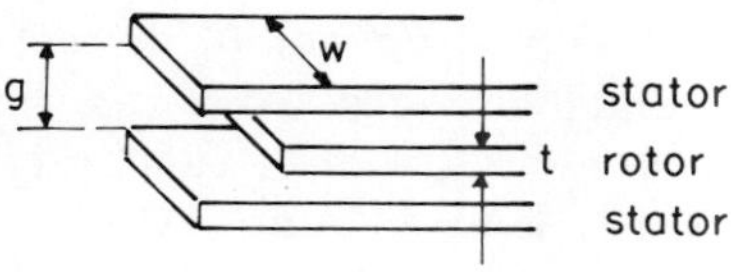

Fig. 7.25.

Application to a short-rotor linear motor gives, (Fig. 7.25):

$$G = \frac{2\mu_0 p^2 f}{\pi\left(\frac{\rho}{t}\right) g \left(\frac{p + w}{w}\right)}$$

allowing for 'end winding resistance' of length $= p$ where p = pole pitch and ρ = resistivity of sheet rotor. Note also that the synchronous speed of the travelling field $= 2pf$, since one cycle corresponds to two pole-pitches. Reference 8 shows that G must be greater than 1 for a single-phase motor to operate.

A linear motor may be operated in an oscillatory mode: (*i*) by reversing the direction of travel of a three-phase field by mechanical switching at each end of rotor travel, (*ii*) in single-phase operation, by springs at the limits of travel to reverse the direction of motion. Note that the single-phase induction motor runs in the preferred direction subject to a start in that direction, i.e. the machine is not self-starting. As stated above the machine must have $G > 1$. (*iii*) by 'back-to-back' operation, i.e. using two three-phase stator windings in opposed directions. The rotor may oscillate subject to, (a) adequate critical speed at the centre, (b) a speed-force curve of suitable form and

(c) $G > 1$, due to analogy between one-phase and three-phase fields.

Other interesting departures from conventional practice are the Arch motor, in which a break is made in the iron of the stator periphery, permitting the rotor flux to decay and be re-established in the break, giving the effect of a non-integral number of pole-pairs. Another way this effect may be achieved is by fitting short-circuited coils in this section.

Problems which arise usually mean a re-appraisal from basic electromagnetics.

Simple theory of a linear device

Consider a linear device with a non-magnetic sheet 'rotor' sandwiched between stator surfaces. A 'current-sheet' distribution is assumed on the airgap surfaces. Also assume, (i) flux penetrates the 'secondary' fully, (i.e. no skin effect, no secondary leakage reactance); (ii) induced current flows straight across the secondary, returning in 'zero-resistance end-bars', (i.e. no transverse end-effects); (iii) induced transient currents in the secondary can be neglected, (i.e. no longitudinal end-effects).

Let $i_s,\ i_r$ = surface distribution of currents, (A/m).

$2p$ = 2(pole-pitch) = wavelength.

Then $$i_s = I_s \sin\left(\omega t - \frac{2\pi x}{2p}\right)$$

being a function of supply frequency, ω and of linear displacement, x. Rotor linear velocity $= V_r$

In elemental terms, for unit width, (Fig. 7.26):

$$\text{Magnetising force} = g\,\frac{\partial H}{\partial x} = (i_r + 2i_s).$$

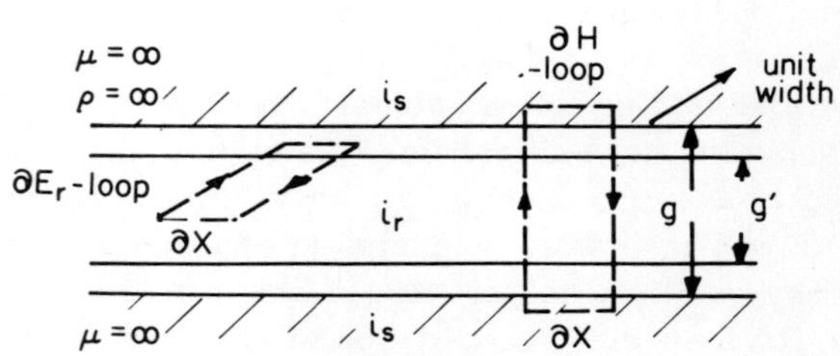

Fig. 7.26.

Also, $$\frac{\partial E_r}{\partial x} = \frac{\partial B}{\partial x}(V_r - 2pf)$$

$$= -2sfp\frac{\partial B}{\partial x} \text{ where } s = \text{slip}$$

$$= -\frac{2sfp\mu_0}{g}(i_r + 2i_s).$$

But, $$\frac{\partial E_r}{\partial x} = \frac{\rho}{g'}\frac{\partial i_r}{\partial x}.$$

Hence, $$\frac{\rho}{g'}\frac{\partial i_r}{\partial x} + \frac{2sfp\mu_0}{g}(i_r + 2i_s) = 0.$$

The solution for i_r will be of the form:

$$i_r = A\sin\left(\omega t - \frac{2\pi x}{2p}\right) + B\cos\left(\omega t - \frac{2\pi x}{2p}\right).$$

With substitution and evaluation of constants we obtain:

$$i_r = \frac{-2I_s sK}{[1 + (sK)^2]}\left\{sK\sin\left(\omega t - \frac{2\pi x}{2p}\right) + \cos\left(\omega t - \frac{2\pi x}{2p}\right)\right\}$$

where $$K = \frac{g'}{g}\frac{2\mu_0 fp^2}{\pi\rho} = \text{'goodness' factor.}$$

The rotor current may be expressed:

$$i_r = \frac{2I_s \cdot sK}{[1 + (sK)^2]^{\frac{1}{2}}}\left\{\sin\left(\omega t - \frac{2\pi x}{2p} + \phi\right)\right\}$$

where $$\cos\phi = \frac{sK}{[1 + (sK)^2]^{\frac{1}{2}}} \quad \text{... (see Fig. 7.27).}$$

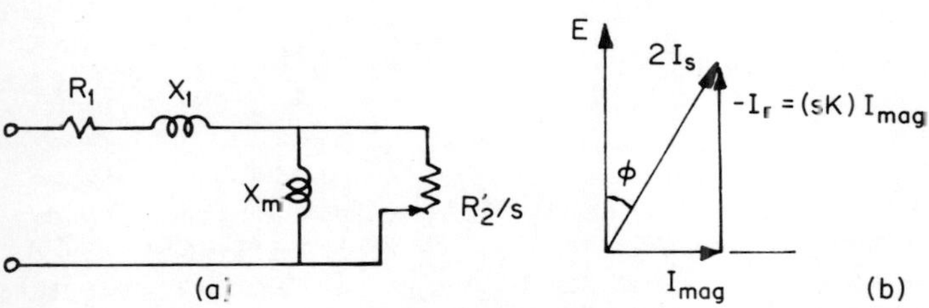

Fig. 7.27.

An expression for the force developed is obtained as follows:

We have, $$\frac{\rho}{g'}\frac{\partial i_r}{\partial x} = -2sfp\frac{\partial B}{\partial x}.$$

Hence $$B = \frac{-\rho i_r}{2sfpg'}$$

$$= \frac{I_s \rho K}{fpg'[1 + (sK)^2]^{\frac{1}{2}}}\left\{\sin\left(\omega t - \frac{2\pi x}{2p} + \phi\right)\right\}.$$

Mean force/unit length $= \frac{1}{2}(B_{\max}\, i_{r\max})$

$$= \frac{1}{2}\;\frac{I_s \rho K}{fpg'[1 + (sK)^2]^{\frac{1}{2}}}\;\frac{2I_s sK}{[1 + (sK)^2]^{\frac{1}{2}}}$$

$$= \frac{I_s^2 sK^2 \rho}{fpg'[1 + (sK)^2]}$$

$$= \frac{2p\mu_0 I_s^2 \cdot sK}{\pi g[1 + (sK)^2]}\ \text{newtons}$$

$$= \frac{2I_s^2 \cdot s}{V_s\left(s^2 + \dfrac{1}{K^2}\right)}\,\frac{\rho}{g'}$$

where $V_s = 2pf =$ synchronous field speed.

N.B.

(a) with constant-current supply, the force peaks when $sK = 1$, or $s = 1/K$,

(b) $sK = I_r/I_{\text{mag}} = \dfrac{X_m}{R_2'/s} = sX_m/R_2'$

that is, $K = X_m/R_2'$.

REFERENCES

1. RAWCLIFFE, G.H. et al, (1958). *Induction motor speed changing by pole amplitude modulation.* Proc. I.E.E. 105A, pp. 411–419.

2. WILLIAMS, F.C. et al, (1964). *Analysis and design of pole change motors using phase mixing techniques.* Proc. I.E.E. III, pp. 80–90.

3. KEMP, P., (1942). *Alternating Current Electrical Engineering*, Macmillan, New York.
4. CLAYTON, A.E., (1921). *Power-Factor Correction*, Pitmans,
5. SCHWARZ, B., (1949). *The Stator-Fed a.c. Commutator Machine with Induction Regulator Control*, Proc. I.E.E. 96, Part II, No. 53, p. 755.
6. KING, K.G. (1965). *Variable Frequency Thyristor Invertors for Induction Motor Speed Control*, Direct Current, 10, p. 26.
7. LAITHWAITE, E.R., (1966). *Induction Machines for Special Purposes*, Newnes,
8. WILLIAMS, F.C. and LAITHWAITE, E.R., (1961). *The Action of the Single-Phase Induction Motor. J.I.E.E.* 7, 551.

EXAMPLES

EXAMPLE 7.1.

Prove that the gross mechanical output of a polyphase induction motor can be represented in the phase-equivalent circuit of the stator by a resistance of value $R_r'(1-s)/s$ in conventional notation.

A three-phase, four-pole, 50 c/s induction motor has a nett output of 14·5 h.p. at 1470 rev/min. If the rotor winding resistance/phase is 0·3 Ω and friction and windage losses are 770 W, calculate the rotor current at this loading.

Solution:

First part. Rotor input $= \omega_s \text{Ⓣ} =$ Rotor o/p + Rotor Cu. loss

$$\text{Rotor output} = \omega_r \text{Ⓣ}$$

$$\therefore \text{Rotor Cu. loss} = (\omega_s - \omega_r)\text{Ⓣ}$$

and
$$\frac{\text{Rotor Cu. loss}}{\text{Rotor input}} = \frac{(\omega_s - \omega_r)}{\omega_s} = \text{slip} = s.$$

Now, $\quad \text{Rotor Cu. loss} = mI_r^2 R_r$ where m = phases.

$$\therefore \quad \text{Rotor input} = m/s\,(I_r^2 R_r).$$

Hence
$$\text{Rotor output} = \left(\frac{m}{s} - m\right) I_r^2 R_r = \left(\frac{1-s}{s}\right) mI_r^2 R_r \text{ watts.}$$

But, Watts in rotor = watts in stator, (when referred).

$$\therefore \quad \left(\frac{1-s}{s}\right) mI_r^2 R_r = \left(\frac{1-s}{s}\right) mI_s^2 R_r' = \text{Rotor o/p}$$

or *per phase* $= I_s^2 R_r'(1-s)/s$ which is equivalent to a resistance of value $R'_r\,(1-s)/s$ in the stator circuit.

Numerical part:

$$\text{Rotor gross output} = (14{\cdot}5 \times 746 + 770)$$
$$= 11\,570$$
$$= \left(\frac{1-s}{s}\right) m I_r^2 R_r$$
$$= \frac{0{\cdot}98}{0{\cdot}02} \times 3 \times I_r^2 \times 0{\cdot}3.$$
$$\text{Hence } I_r^2 = 262$$
$$\text{and } I_r = 16{\cdot}2\ \text{A}.$$

EXAMPLE 7.2.

A six-pole 50 c/s three-phase induction motor operating on 415 V a.c. mains has a power factor of 0·75 lagging when developing a useful torque of 120 lb ft at a slip frequency of two c/s. The mechanical torque lost in friction is 10 lb ft and the stator losses equal the rotor copper losses. Calculate (a) the b.h.p., (b) the line current, and (c) the efficiency.

Solution:

(a) Running frequency, $f = 48\ \text{c/s} = pn/60$

$$\therefore \quad \text{Speed, } n = 60 \times 48/3 = 960\ \text{r.p.m.}$$

$$\text{b.h.p.} = \frac{2\pi n\ \textcircled{T}}{33\,000} = \frac{2\pi \times 960 \times 120}{33\,000} = 22\ \text{h.p.}$$

(b) $$\text{Rotor intake} = 2\pi n_s\ \textcircled{T}_g \times \frac{746}{33\,000} \quad \text{where } \textcircled{T}_g = \text{gross torque}$$

$$= \frac{2\pi \times 1000 \times (120 + 10) \times 746}{33\,000}$$
$$= 18\,400\ \text{W}.$$

$$\text{Rotor cu. loss} = \text{slip} \times \text{rotor intake}$$
$$= \frac{2}{50} \times 18\,400 = 736\ \text{W}$$
$$= \text{stator losses, (given).}$$

$$\therefore \quad \text{Total input} = 736 + 18\,400 = 19\,136\ \text{W}.$$

$$\therefore \quad \text{Efficiency, } \eta = \frac{22 \times 746}{19\,136} = 0{\cdot}86 \text{ p.u.}$$

(c)
$$\text{Input} = \sqrt{3}\, V_L I_L \cos\phi = 19\,136 \text{ W}$$

$$\therefore \quad I_L = \frac{19\,136}{\sqrt{3} \times 415 \times 0{\cdot}75} = 35{\cdot}3 \text{ A.}$$

EXAMPLE 7.3.

A 25 h.p. six-pole 50 c/s three-phase induction motor runs at 960 r.p.m. on full load, with a rotor current of 35 amps/phase. Allowing 250 W for copper loss in the short-circuited starter and 1000 W for mechanical losses, calculate the resistance/phase of the rotor winding.

Solution:

$$\text{Output} = 25 \text{ h.p.} = 25 \times 746 = 18\,600 \text{ W}$$

$$\text{Mechanical loss} = 1000 \text{ W, (given)}$$

$$\therefore \text{Input to rotor} = 19\,600 + \text{Rotor cu. loss.}$$

$$\text{Now,} \quad \text{slip} = \frac{1000 - 960}{1000} = 0{\cdot}04.$$

$$\text{Then, Rotor cu. loss} = 0{\cdot}04\,(19\,600 + \text{rotor cu. loss})$$

$$0{\cdot}96 \text{ (rotor cu. loss)} = 784$$

$$\therefore \quad \text{rotor cu. loss} = 820 \text{ W}$$

$$\text{Then, rotor winding cu. loss} = 820 - 250 = 570 \text{ W}$$

$$= \frac{570}{3} \doteq 190 \text{ W/phase.}$$

$$\text{Hence, rotor winding resistance/phase} = \frac{190}{35^2} = 0{\cdot}156\Omega.$$

EXAMPLE 7.4.

A given polyphase induction motor operates on constant-voltage, constant-frequency mains. Using an analysis which neglects stator impedance, (*i*) determine the dependance of torque on slip, s and the ratio, a, where:

$$a = \frac{\text{rotor resistance/ph.}}{\text{rotor standstill reactance/ph.}}$$

(*ii*) draw curves to show the relationship torque/slip and power/slip, for ratios $a = 0{\cdot}1$ and $a = 1{\cdot}0$, (*iii*) calculate the external rotor resistance/phase required to give 2/3 $Ⓣ_{max}$ at starting, given:

rotor winding resistance/phase $= 0{\cdot}02\,\Omega$

rotor standstill reactance/phase $= 0{\cdot}1\,\Omega$.

Solution:

(*i*) $a = R_r/\omega L_r \quad \therefore \quad R_r = a(\omega L_r)$

From the elementary analysis, (p. 219):

$$Ⓣ = \frac{m}{\omega_s k^2} \cdot \frac{sV^2 R_r}{R_r^2 + X_r^2}$$

$$\propto \frac{sR_r}{R_r^2 + (s\omega L_r)^2} \propto \frac{sa(\omega L_r)}{(\omega L_r)^2\,(a^2 + s^2)}.$$

Now ωL_r = constant. $\therefore \quad Ⓣ \propto \dfrac{sa}{(a^2 + s^2)}$.

Note also, speed $n_r = n_s(1 - s)$

power, $P \propto n_r\,Ⓣ \propto (1 - s)\,Ⓣ$.

(*ii*) Taking a range of values for s, we have in arbitary units:

$a = 0{\cdot}1$				$a = 1{\cdot}0$			
s	n_r	Ⓣ	P	s	n_r	Ⓣ	P
1·0	0	0·099	0	1·0	0	0·5	0
0·8	0·2	0·123	0·0246	0·8	0·2	0·487	0·0974
0·6	0·4	0·162	0·0648	0·6	0·4	0·441	0·1764
0·4	0·6	0·235	0·141	0·4	0·6	0·344	0·2064
0·2	0·8	0·4	0·32	0·2	0·8	0·192	0·1536
0·1	0·9	0·5	0·45	0	1·0	0	0
0	1·0	0	0				

These results are shown plotted in Fig. Ex. 7.4.

(*iii*) At maximum torque, $s = \dfrac{R_r}{\omega L_r} = \dfrac{0{\cdot}02}{0{\cdot}1} = 0{\cdot}2.$

Now $$\text{Torque} \propto \frac{sR_r}{R_r^2 + (s\omega L_r)^2}$$

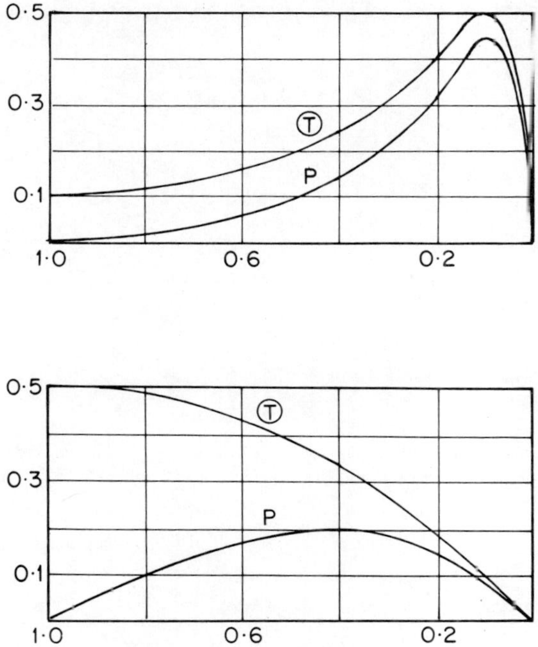

Fig. Ex. 7.4.

$$\therefore \quad \textcircled{T}_{max} \propto \frac{0 \cdot 2 \times 0 \cdot 02}{(0 \cdot 02)^2 + (0 \cdot 2 \times 0 \cdot 1)^2}$$

$$= \frac{0 \cdot 2}{0 \cdot 04} = 5 \text{ arbitary units.}$$

$$\frac{2}{3}\,\textcircled{T}_{max} = \frac{10}{3} \text{ arbitary units.}$$

At start, $s = 1 \cdot 0$ and added rotor resistance $= R$,

$$\therefore \quad \textcircled{T}_{start} \propto \frac{(R_r + R)}{(R_r + R)^2 + (\omega L_r)^2}$$

$$\frac{10}{3} = \frac{(0 \cdot 02 + R)}{(0 \cdot 02 - R)^2 + (0 \cdot 1)^2} = \frac{(0 \cdot 02 + R)}{R^2 + 0 \cdot 04R + 0 \cdot 0104}$$

$$10\,R^2 - 2 \cdot 6\,R + 0 \cdot 044 = 0$$

$$\therefore \qquad R = \frac{2{\cdot}6 \pm (6{\cdot}76 - 1{\cdot}76)^{\frac{1}{2}}}{20}$$

$$= 0{\cdot}242 \text{ or } 0{\cdot}018\,\Omega$$

8 Introduction to generalised machine theory

Generalised machine theory

Most electrical machines can be interpreted in terms of a number of interacting coupled coils. Equations governing voltages, currents and torques for the general case, where the relative disposition of the coils was random, were obtained, (p. 62), as follows:

$$[e] = \{[r] + p[L] + \omega\left[\frac{\mathrm{d}L}{\mathrm{d}\theta}\right]\}[i]$$

$$[\text{Ⓣ}] = \frac{1}{2}[i_t]\left[\frac{\mathrm{d}L}{\mathrm{d}\theta}\right][i].$$

Now in practice the active coils of machines are located on precise axes, for example, in d.c. machines the principal coils are located on the direct and quadrature axes. But in a three-phase machine the principal coils are oriented on axes spaced at 120°, in the case of a two-pole system, and the field coil varies with respect to the fixed axes in terms of $\mathrm{d}\theta/\mathrm{d}t$. Functions relating to any axis can be transformed, by mathematical resolution, into terms relating to direct and quadrature axes.

Hence the concept of a two-axis two-pole 'primitive' machine arises, the properties of which can form the basis of a unified theory for machines.

The primitive machine

The primitive machine chosen for unified theory is endowed with properties which bear direct relationship to real machines. The concept of a rotating part and a stationary part is retained, but in a special way.

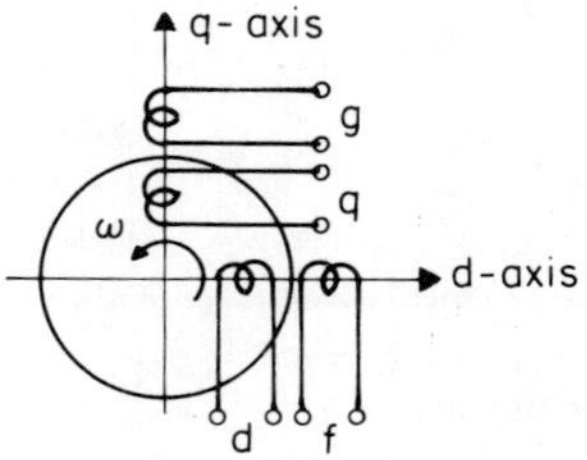

Fig. 8.1.

The four coils shown define the possible positions in which a coil may exist on either axis, on both the stator and the rotor, the rotor being shown as the inner member, Fig. 8.1. Each coil may be subject to a self-induced e.m.f., a transformer e.m.f. and a rotational e.m.f., as suggested by the general e.m.f. equation above. However, the choice of a two-axis primitive machine is deliberate and this imposes certain idealised conditions on the inter-relationship of the coils as follows:

Linear magnetisation and a sinusoidally distributed airgap m.m.f. are assumed so that direct and quadrature axis sinusoidal m.m.f.'s can be resolved. The coils are stationary on the axes as shown. Then:

(a) coils on the same axis have mutual coupling. Current change in one coil induces a transformer e.m.f. in other coils on the same axis.

(b) a coil on one axis has no transformer coupling with a coil on the other axis.

(c) the fixed coils shown on the rotor part are called 'pseudo-stationary' to justify the rotational voltage induced in them, normally acquired by rotation in flux on the next axis in the defined direction of rotation.

(d) voltage impressed on a coil = e, supply current to a coil in the direction of e = i. Positive current to a coil produces a flux polarity according to the convention shown in Fig. 1.1.

Hence for the four coils shown:

$$e_f = (r_f + L_f p)\, i_f + L_{fd}\, p i_d$$
$$e_d = (r_d + L_d p)\, i_d + L_{df}\, p i_f + \omega G_{dq} i_q + \omega G_{dg} i_g$$
$$e_q = (r_q + L_q p)\, i_q + L_{qg}\, p i_g + \omega G_{qf}\, i_f + \omega G_{qd}\, i_d$$
$$e_g = (r_g + L_g p)\, i_g + L_{gq}\, p i_q$$

Now in the Kron definition of a primitive machine all input quantities are positive. Thus the Kron machine must be recognised only as a mathematical model defining positive quantities, for example, with positive voltages and currents in a 'real' machine, (i.e. motoring operation), the speed will be clockwise, Fig. 8.2. Relative to the Kron model the motor will develop negative speed, since speed is an output quantity. Thus the definition of positive speed is anticlockwise.

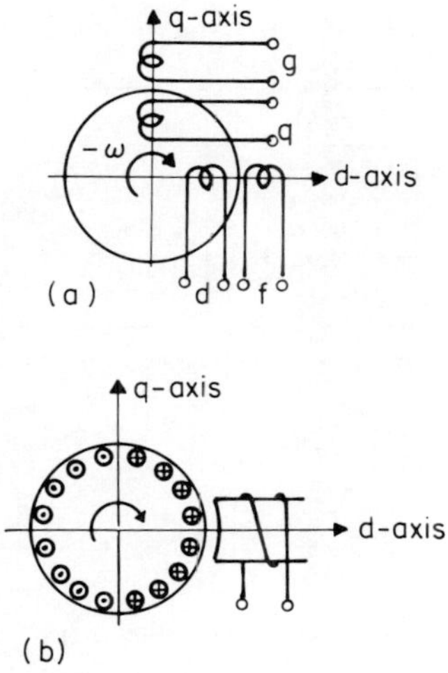

Fig. 8.2.

The G coefficients are 'rotational' inductance coefficients which arise due to mutual inductance which occurs when a coil has been advanced $\pi/2$ to align with the axis of the induction flux. Now since $M = M_{\max} \cos\theta$ and $G = \mathrm{d}M/\mathrm{d}\theta = M_{\max}$ at $\theta = \pi/2$, then for positive rotation:

$$G_{qf} = -M_d$$
$$G_{dg} = M_q.$$

The d and q coils are assumed to be the same set of rotor conductors, which produce two independant fluxes along the designated fixed axes of different reluctances. Hence:

$$G_{dq} = L_q$$

$$G_{qd} = -L_d$$

Also $$L_{df} = M_d \text{ and } L_{qg} = M_q$$

The e.m.f. equations for the Kron primitive machine may now be expressed:

$$\begin{bmatrix} e_f \\ e_d \\ e_q \\ e_g \end{bmatrix} = \begin{bmatrix} r_f + L_f p & M_d p & & \\ M_d p & r_d + L_d p & \omega L_q & \omega M_q \\ -\omega M_d & -\omega L_d & r_q + L_q p & M_q p \\ & & M_q p & r_g + L_g p \end{bmatrix} \begin{bmatrix} i_f \\ i_d \\ i_q \\ i_g \end{bmatrix}$$

The e.m.f. equations for a primitive 'motor' will then be:

$$\begin{bmatrix} e_f \\ e_d \\ e_q \\ e_g \end{bmatrix} = \begin{bmatrix} r_f + L_f p & M_d p & & \\ M_d p & r_d + L_d p & -\omega L_q & -\omega M_q \\ \omega M_d & \omega L_d & r_q + L_q p & M_q p \\ & & M_q p & r_g + L_g p \end{bmatrix} \begin{bmatrix} i_f \\ i_d \\ i_q \\ i_g \end{bmatrix}$$

The power supplied to the motor:

$$P = \sum ei$$

$$= \left[i_t ri + i_t L \frac{di}{dt} + P_e \right].$$

The power developed, of electrical origin:

$$P_e = \omega(Jp\omega + D\omega) + \omega \textcircled{T}_s$$

and $$\textcircled{T}_e = \frac{P_e}{\omega} = Jp\omega + D\omega + \textcircled{T}_s\,.$$

From the matrix for the motor:

$$\begin{aligned} P_e &= e_d i_d + e_q i_q \\ &= i_d(-\omega L_q i_q - \omega M_q i_g) + i_q(\omega M_d i_f + \omega L_d i_d) \\ &= \omega\{(M_d i_f i_q + L_d i_d i_q) - (L_q i_q i_d + M_q i_g i_d)\} \end{aligned}$$

$$\therefore \quad Ⓣ_e = \frac{P_e}{\omega} = M_d i_f i_q - M_q i_g i_d + i_d i_q (L_d - L_q).$$

Operating as a generator:

$$P_s = \omega Ⓣ_s = \omega\{(Jp\omega + D\omega) + Ⓣ_e\}$$

or $$Ⓣ_s = Jp\omega + D\omega + Ⓣ_e .$$

If the machine has a uniform airgap, $L_d = L_q$ and the term $i_d i_q (L_d - L_q)$ disappears; if $L_d \neq L_q$ the term remains and the machine develops reluctance torque.

The above equilibrium equations for primitive machines have direct application to d.c. machines since d.c. machines are already two-axis machines. Application to a.c. machines requires that the a.c. machine quantities be referred, or transformed into terms of d and q axes, as shown below.

Approximate solutions are obtained because of the assumed approximations of sinusoidal flux distribution and linear magnetisation. The matrix may be adjusted by removing or adding rows and columns, appropriate to the actual windings which occur in the real machine.

General equations with stationary axes

The above equations, with reference to stationary axes, may be expressed in general form:

$$[e] = [ir + Lpi + \omega Gi]$$

and $$[e]\,[i_t] = [i_t ri + i_t Lpi + i_t \omega Gi]$$

where the terms on the right hand side define copper loss, power stored magnetically and power developed as torque respectively. Thus torque $= [i_t Gi]$.

It follows that $[i_t Gi]$ must $= \frac{1}{2}\left[i_t \frac{\mathrm{d}L}{\mathrm{d}\theta} i\right]$.

To show that $\frac{1}{2}\left[i_t \frac{\mathrm{d}L}{\mathrm{d}\theta} i\right] = [i_t Gi]$: (Ref. 5).

Consider a simple reluctance motor having 2 coils at 90° located on the rotor, Fig. 8.3. Then:

$$\text{Reluctance torque, } Ⓣ_r = \frac{1}{2} i_a^2 \frac{\mathrm{d}L_a}{\mathrm{d}\theta} + \frac{1}{2} i_b^2 \frac{\mathrm{d}L_b}{\mathrm{d}\theta} .$$

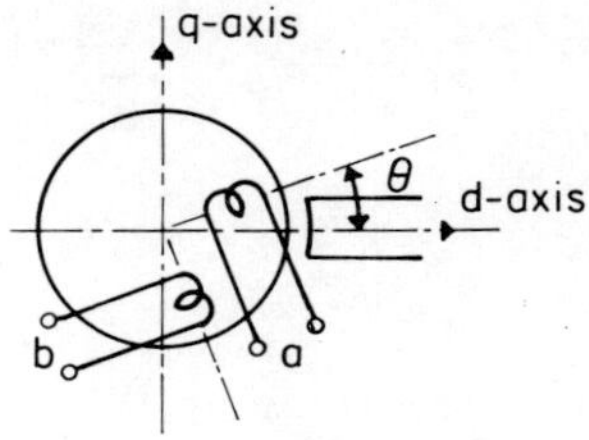

Fig. 8.3.

Transforming the coil currents to fictitious coils on stationary $d-q$ axes:

$$i_d = i_a \cos\theta + i_b \sin\theta$$
$$i_q = i_a \sin\theta - i_q \cos\theta.$$

Also,

$$\lambda_d = L_d i_d = L_d(i_a \cos\theta + i_b \sin\theta) = \lambda_{da} + \lambda_{db}$$
$$\lambda_q = L_q i_q = L_q(i_a \sin\theta - i_b \cos\theta) = \lambda_{qa} + \lambda_{qb}.$$

Now

$$\lambda_a = \lambda_{da} \cos\theta + \lambda_{qa} \sin\theta$$
$$\lambda_b = \lambda_{db} \sin\theta - \lambda_{qb} \cos\theta$$
$$\therefore \quad \lambda_a = L_d(i_a \cos^2\theta) + L_q(i_a \sin^2\theta) = L_a i_a$$
$$L_a = L_d \cos^2\theta + L_q \sin^2\theta$$
$$= L_d\left(\frac{1 + \cos 2\theta}{2}\right) + L_q\left(\frac{1 - \cos 2\theta}{2}\right)$$
$$= \frac{L_d + L_q}{2} + \frac{L_d - L_q}{2} \cos 2\theta.$$

Similarly:

$$L_b = \frac{L_d + L_q}{2} - \frac{L_d - L_q}{2} \cos 2\theta$$

$$\frac{dL_a}{d\theta} = -(L_d - L_q) \sin 2\theta$$

$$\frac{dL_b}{d\theta} = (L_d - L_q) \sin 2\theta.$$

Using a reverse transformation for currents:

$$i_a = i_d \cos\theta + i_q \sin\theta$$
$$i_b = i_d \sin\theta - i_q \cos\theta.$$

Then

$$i_a^2 = i_d^2 \cos^2\theta + i_q^2 \sin^2\theta + 2 i_d i_q \sin\theta\cos\theta$$
$$i_b^2 = i_d^2 \sin^2\theta + i_q^2 \cos^2\theta - 2 i_d i_q \sin\theta\cos\theta.$$

Substituting in the above expression for torque:

$$Ⓣ_r = \tfrac{1}{2}\{[i_d^2 \cos^2\theta + i_q^2 \sin^2\theta + 2 i_d i_q \sin\theta\cos\theta][-(L_d - L_q)\sin 2\theta]$$
$$+ [i_d^2 \cos^2\theta + i_q^2 \sin^2\theta - 2 i_d i_q \sin\theta\cos\theta][(L_d - L_q)\sin 2\theta]\}$$
$$= - i_d i_q (L_d - L_q).$$

Alternatively, we may determine the reluctance torque:

$$Ⓣ_r = [i_t\, Gi] = \begin{vmatrix} i_d & i_q \end{vmatrix} \begin{bmatrix} & L_q \\ -L_d & \end{bmatrix} \times \begin{vmatrix} i_d \\ i_q \end{vmatrix}$$

$$= i_d i_q L_q - i_q i_d L_d$$
$$= - i_d i_q (L_d - L_q).$$

The proof also demonstrates that:

$$[G] = \begin{bmatrix} & L_q \\ -L_d & \end{bmatrix} = \frac{1}{2}\begin{bmatrix} & -(L_d - L_q) \\ -(L_d - L_q) & \end{bmatrix} + \frac{1}{2}\begin{bmatrix} & (L_d + L_q) \\ -(L_d + L_q) & \end{bmatrix}$$

$$= G_{\text{sym}} + G_{\text{skew}}$$

and we note that a skew symmetric matrix is always zero.

The d.c. machine

Normally the machine has a field coil on the d-axis and, with brushes located on the q-axis, a rotor q-coil. Hence the general statement of e.m.f. equations reduces to:

$$\begin{bmatrix} e_f \\ e_q \end{bmatrix} = \begin{bmatrix} r_f + L_f p & \\ \omega M_d & r_q + L_q p \end{bmatrix} \times \begin{bmatrix} i_f \\ i_q \end{bmatrix}$$

or

$$e_f = i_f(r_f + L_f p)$$
$$e_q = \omega M_d i_f + i_q(r_q + L_q p)$$

and

$$Ⓣ_e = i_q i_f M_d.$$

Under conditions of steady-state operation, ω = constant speed and $p = 0$ since there is no rate of change of currents. Thus the equations for steady-state operation are:

$$e_f = I_f r_f$$
$$e_q = \omega M_d I_f + I_q r_q$$
$$Ⓣ_e = I_q I_f M_d.$$

These results naturally agree with the equations used in Chapter 5, where the same basic assumptions were used to obtain linearised equations and solutions.

The d.c. machine with series field windings

Modern machines may have additional windings such as interpole and compensating windings, located on the q-axis. To allow for these the appropriate (g) column and row could be reinstated in the general matrix. However, since they are in series with the armature they may be combined with the q-coil, and this treated as a single armature winding but with amended values of inductance and resistance.

In compound-wound machines, a series winding is introduced on the direct axis of the stator, *coil s*, and the voltage equations become:

$$e_f = (r_f + L_f p)\, i_f + M_{fs} p i_a$$
$$e_a = \omega M_f\, i_f + \omega N M_f\, i_a + (r_a + L_a p)\, i_a + M_{fs} p i_f$$

also

$$Ⓣ_e = (M_f\, i_f + N M_f\, i_a)\, i_a$$

where

$$N = \text{ratio } \frac{\text{turns in coil } s}{\text{turns in coil } f}.$$

Generalised a.c. machine equations

It is conventional to represent the three-phase stator winding of an a.c. machine, in idealised form, by three phase coils on the inner member or rotor. The voltage impressed on each phase coil can be expressed in the form:

$$e = ir + \frac{d\lambda}{dt}$$

where λ = total flux-linkages with the phase coil. The self and mutual inductance terms involved in λ will vary with θ, the instantaneous position of the coil concerned. The complicated expressions which arise lead to unweildly algebra in the solution of machine problems. However, by transforming the phase quantities by mathematical resolution, a two-axis model can be derived which conforms to the Kron primitive machine.

Transformations

We can determine m.m.f's $\mathfrak{F}_d$ and $\mathfrak{F}_q$ which will produce a resultant m.m.f. equivalent to that produced by $\mathfrak{F}_a$, $\mathfrak{F}_b$ and $\mathfrak{F}_c$, (Fig. 8.4(a)) by mathematical resolution, thus:

$$\mathfrak{F}_d = \mathfrak{F}_a \cos\theta + \mathfrak{F}_b \cos\left(\theta - \frac{2\pi}{3}\right) + \mathfrak{F}_c \cos\left(\theta - \frac{4\pi}{3}\right).$$

Similarly:

$$\mathfrak{F}_q = \mathfrak{F}_a \sin\theta + \mathfrak{F}_b \sin\left(\theta - \frac{2\pi}{3}\right) + \mathfrak{F}_c \sin\left(\theta - \frac{4\pi}{3}\right).$$

A decision has to be made concerning the number of turns in the coils. If the d and q coils are defined as having the same number of turns as a phase coil, then since a three-phase system produces a resultant flux $3/2\ \Phi_m$ and a two-phase system Φ_m, for flux equality a 2/3 factor must be introduced into the current transformation, viz:

$$i_d = \frac{2}{3}\left[i_a \cos\theta + i_b \cos\left(\theta - \frac{2\pi}{3}\right) + i_c \cos\left(\theta - \frac{4\pi}{3}\right)\right]$$

$$i_q = \frac{2}{3}\left[i_a \sin\theta + i_b \sin\left(\theta - \frac{2\pi}{3}\right) + i_c \sin\left(\theta - \frac{4\pi}{3}\right)\right].$$

If written in matrix form, we note that the matrix is non-square, and prevents derivation of (abc) terms from (dq) terms. Alternatively, we note that there is a unique pair of currents, i_d and i_q, to produce a given m.m.f., but there is no unique set of three-phase currents i_a, i_b and i_c, viz: Suppose a current i_0 is passed through the three coils in series. The resultant flux is zero. Hence, to each of the three-phase currents i_a, i_b and i_c we could add i_0 and not change the flux pattern. This added current component is termed the zero-sequence component. In symmetrical-component analysis it is determined by adding i_a, i_b and i_c vectorially and dividing by 3, or $i_0 = 1/3\,(i_a + i_b + i_c)$. Although i_0 produces no resultant flux it will cause leakage. Generally we shall be dealing with balanced conditions.

By including zero-sequence current we obtain a square matrix for currents:

$$\begin{vmatrix} i_d \\ i_q \\ i_0 \end{vmatrix} = \frac{2}{3}\begin{vmatrix} \cos\theta & \cos\left(\theta - \dfrac{2\pi}{3}\right) & \cos\left(\theta - \dfrac{4\pi}{3}\right) \\ \sin\theta & \sin\left(\theta - \dfrac{2\pi}{3}\right) & \sin\left(\theta - \dfrac{4\pi}{3}\right) \\ \dfrac{1}{2} & \dfrac{1}{2} & \dfrac{1}{2} \end{vmatrix} \times \begin{vmatrix} i_a \\ i_b \\ i_c \end{vmatrix}$$

The inverse transformation gives values of i_a, i_b and i_c in terms of i_d, i_q and i_0 as follows:

$$\begin{vmatrix} i_a \\ i_b \\ i_c \end{vmatrix} = \begin{vmatrix} \cos\theta & \sin\theta & 1 \\ \cos\left(\theta - \dfrac{2\pi}{3}\right) & \sin\left(\theta - \dfrac{2\pi}{3}\right) & 1 \\ \cos\left(\theta - \dfrac{4\pi}{3}\right) & \sin\left(\theta - \dfrac{4\pi}{3}\right) & 1 \end{vmatrix} \times \begin{vmatrix} i_d \\ i_q \\ i_0 \end{vmatrix}$$

The voltage transformations are defined by a similar set of equations as above, except that i is replaced by e.

Unit power, defined on a per-unit basis, occurs when there is unit current and unit voltage in each main circuit, or:

$$P = k_p \Sigma\, ei, \text{ where } k_p = \frac{1}{\text{number of main circuits}}\,.$$

Hence for a three-phase machine:

$$P = \tfrac{1}{3}(e_a i_a + e_b i_b + e_c i_c).$$

Substituting the above transformations in this expression, we obtain:

$$P = \tfrac{1}{2}(e_d i_d + e_q i_q) + e_0 i_0$$

and with balanced loading $e_0 i_0$ is zero.

Alternative forms of the transformation matrix

Further unification of the transformation expressions and a simplification of the mathematics involved, is obtained as described below. The transformation, C, required involves two basic transformations, commonly termed the 'Phase transformation', viz. C_{012}^{abc}, (or simply C_2^a), which transforms from three-phase to two-phase synchronous axes, and the 'Commutator transformation', C_{odq}^{012}, (or simply C_d^2), which removes the θ-dependant property and introduces the pseudo-stationary property for certain coils on fixed dq axes.

Consider the C_2^a transformation:

$$I_{012} = [C_2^a]\, I_{abc} \quad \text{and hence} \quad I_{abc} = [C_2^a]^{-1}\, I_{012}$$

$$E_{012} = [C_2^a]\, E_{abc} \quad \text{and hence} \quad E_{abc} = [C_2^a]^{-1}\, E_{012}\,.$$

Now, if $I_{abc} = [C_{2t}^a]\, I_{012}$ where $[C_{2t}^a]$ is the transpose of $[C_2^a]$, then $[C_{2t}^a][C_2^a] = 1$, (unit matrix), and the transformation is termed 'orthogonal'.

To obtain this condition we can say; Fig. 8.4 (a) and (b):

$$\begin{vmatrix} i_0 \\ i_1 \\ i_2 \end{vmatrix} = k_1 x \begin{vmatrix} k & k & k \\ 1 & -\frac{1}{2} & -\frac{1}{2} \\ 0 & \frac{\sqrt{3}}{2} & -\frac{\sqrt{3}}{2} \end{vmatrix} \times \begin{vmatrix} i_a \\ i_b \\ i_c \end{vmatrix}$$

or more concisely, $\quad I_{012} = [C_2^a]\, I_{abc}\,.$

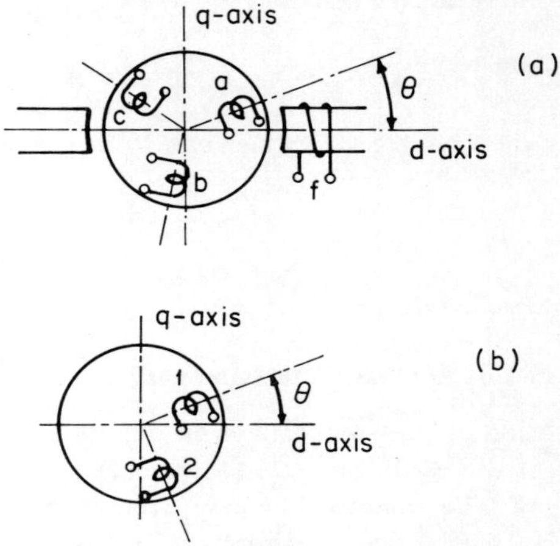

Fig. 8.4.

Hence $$[C_2^a] = k_1 \times \begin{vmatrix} k & k & k \\ 1 & -\frac{1}{2} & -\frac{1}{2} \\ 0 & \frac{\sqrt{3}}{2} & -\frac{\sqrt{3}}{2} \end{vmatrix}$$

and $$[C_{2t}^a] = k_1 \times \begin{vmatrix} k & 1 & 0 \\ k & -\frac{1}{2} & \frac{\sqrt{3}}{2} \\ k & -\frac{1}{2} & -\frac{\sqrt{3}}{2} \end{vmatrix}$$

Then $$[C_2^a]\,[C_{2t}^a] = k_1^2 \times \begin{vmatrix} 3k^2 & 0 & 0 \\ 0 & \frac{3}{2} & 0 \\ 0 & 0 & \frac{3}{2} \end{vmatrix}$$

This is a 'unit matrix' if $3k_1^2k^2 = 1$ and $(3/2)\,k^2 = 1$ whence $k_1 = \sqrt{(2/3)}$ and $k = \sqrt{(1/2)}$.

Substituting in the expressions above we obtain the desired values for C_2^a and C_{2t}^a viz:

$$[C_2^c] = \sqrt{\frac{2}{3}} \times \left| \begin{array}{ccc} \frac{1}{\sqrt{2}} & \frac{1}{\sqrt{2}} & \frac{1}{\sqrt{2}} \\ 1 & -\frac{1}{2} & -\frac{1}{2} \\ 0 & \frac{\sqrt{3}}{2} & -\frac{\sqrt{3}}{2} \end{array} \right|$$

and $$[C_{2t}^a] = \sqrt{\frac{2}{3}} \times \left| \begin{array}{ccc} \frac{1}{\sqrt{2}} & 1 & 0 \\ \frac{1}{\sqrt{2}} & -\frac{1}{2} & \frac{\sqrt{3}}{2} \\ \frac{1}{\sqrt{2}} & -\frac{1}{2} & -\frac{\sqrt{3}}{2} \end{array} \right|$$

Using the relations, $I_{012} = [C_2^a]\,I_{abc}$ and $E_{012} = [C_2^a]\,E_{abc}$ we obtain:

$$i_0 = \frac{1}{\sqrt{3}}\,(i_a + i_b + i_c)$$

$$i_1 = \sqrt{\frac{2}{3}}\,(i_a - \frac{1}{2}\,i_b - \frac{1}{2}\,i_c)$$

$$i_2 = \frac{1}{\sqrt{2}}\,(i_b - i_c).$$

Identical relations are obtained for e_0, e_1, e_2 in terms of e_a, e_b and e_c.

The commutator transformation, C_{odq}^{012}, (or simply C_d^2) is obtained by trignometrical resolution, (Fig. 8.4(b)), and the zero-sequence component is included because we shall wish to combine C_d^2 with C_2^a to obtain the overall transformation C_{odq}^{abc}, (referred to latterly simply as C).

Now

$$[C_d^2] = \begin{vmatrix} 1 & 0 & 0 \\ 0 & \cos\theta & \sin\theta \\ 0 & \sin\theta & -\cos\theta \end{vmatrix}.$$

The inverse transformation C_2^d proves to be identical (due to the arbitary choice of the q-axis leading the d-axis).

The complete transformation, C, from abc axes to odq fixed axes is thus obtained:

$$[C] = [C_2^a][C_d^2] = \sqrt{\frac{2}{3}} \begin{vmatrix} \frac{1}{\sqrt{2}} & \frac{1}{\sqrt{2}} & \frac{1}{\sqrt{2}} \\ \cos\theta & \cos\left(\theta - \frac{2\pi}{3}\right) & \cos\left(\theta - \frac{4\pi}{3}\right) \\ \sin\theta & \sin\left(\theta - \frac{2\pi}{3}\right) & \sin\left(\theta - \frac{4\pi}{3}\right) \end{vmatrix}$$

$$[C_t] = \sqrt{\frac{2}{3}} \begin{vmatrix} \frac{1}{\sqrt{2}} & \cos\theta & \sin\theta \\ \frac{1}{\sqrt{2}} & \cos\left(\theta - \frac{2\pi}{3}\right) & \sin\left(\theta - \frac{2\pi}{3}\right) \\ \frac{1}{\sqrt{2}} & \cos\left(\theta - \frac{4\pi}{3}\right) & \sin\left(\theta - \frac{4\pi}{3}\right) \end{vmatrix}.$$

These relations can be obtained directly from the m.m.f. relation on p. 255 if it is recognised that:

(a) This latter transformation depends on an alternative choice of turns ratio for the coils, viz:

$$N_a/N_2 = N_a/N_d = \sqrt{(2/3)} \text{ and } N_2/N_d = 1.$$

(b) We can write
$$\mathcal{F}_{012} = N_2\, i_{012}$$
$$= N_2\, [C_2^a]\, i_{abc}.$$

Now with $N_a/N_2 = \sqrt{(2/3)}$

$$\therefore \quad \mathcal{F}_{012} = \sqrt{\frac{3}{2}}\,[C_2^a]\, i_{abc} N_a = \sqrt{\frac{3}{2}}\,[C_2^a]\, \mathcal{F}_{abc}.$$

Also, $N_2/N_d = 1$

$$\therefore \quad \mathcal{F}_{odq} = \sqrt{\frac{3}{2}}\,[C_d^a]\, \mathcal{F}_{abc}$$

and

$$\mathcal{F}_{abc} = \sqrt{\frac{2}{3}}\,[C_{dt}^a]\, \mathcal{F}_{odq}.$$

It should also be noted that because $[C][C_t]$ = unit matrix:

$$[e_{odq}]_t\,[i_{odq}] = [e_{abc}]_t\,[i_{abc}]$$

and the $[C]$ transformations are termed Power Invariant since the sum of the instantaneous powers of the *abc* machine is equal to that for the *dq* model.

Inductance coefficients

Field self-inductance, L_f

The magnetic circuit on the field coil axis is undisturbed by rotation, except for slotting which is neglected. Hence L_f remains constant.

Mutual inductances between phase coils, L_{ab} etc.

Assuming that coils on the same axis have coupling and coils at $\pi/2$ have no coupling, the component of *a*-axis coupling between coils *a* and $b = L_1 \cos 2\pi/3 = -1/2\, L_1$.

Now introduction of the stator iron gives a θ-dependant variation, [Fig. 8.4 (a)],

$$\therefore \quad L_{ab} = -\frac{1}{2} L_1 + L_2 \cos 2\theta$$

$$L_{bc} = -\frac{1}{2} L_1 + L_2 \cos\left(2\theta - \frac{2\pi}{3}\right)$$

$$L_{ca} = -\frac{1}{2} L_1 + L_2 \cos\left(2\theta - \frac{4\pi}{3}\right).$$

Self-inductance of a phase coil, L_a etc.

Including leakage inductance and the θ-dependant variation due to stator iron:

$$L_a = L_l + L_1 + L_2 \cos 2\theta$$

$$L_b = L_l + L_1 + L_2 \cos\left(2\theta - \frac{2\pi}{3}\right)$$

$$L_c = L_l + L_1 + L_2 \cos\left(2\theta - \frac{4\pi}{3}\right).$$

Phase coil- field coil mutual inductance, L_{af}, etc.
If M = the maximum (co-axial) mutual inductance between a phase coil and the field coil:

$$L_{af} = M\cos\theta$$

$$L_{bf} = M\cos\left(\theta - \frac{2\pi}{3}\right)$$

$$L_{cf} = M\cos\left(\theta - \frac{4\pi}{3}\right).$$

Voltage and impedance relationships

Using the abbreviated sub and superscripts on *all* terms, we have:

$$i_d = [C_d^a]\, i_a \quad \text{and} \quad i_a = [C_{dt}^a]\, i_d$$

or more simply:

$$i_d = [C]\, i_a \quad \text{and} \quad i_a = [C_t]\, i_d\,.$$

Also: $$e_d = [C]\, e_a \quad \text{and} \quad e_a = [C_t]\, e_d.$$

For the phase coils we can state:

$$e_a = i_a r_a + p(L i_a)$$

where L will be defined below.

Then
$$\begin{aligned}
e_a &= r_a [C_t]\, i_d + p\,\{L[C_t]\, i_d\} \\
e_d &= [C]\, e_a \\
&= [C]\, r_a\, [C_t]\, i_d + [C]\, p\,\{L\,[C_t]\, i_d\} \\
&= r_d i_d + [C]\, L\,[C_t]\, p i_d + i_d [C]\, p\,\{L\,[C_t]\} \\
&= i_d r_d + L_d p i_d + \omega G i_d.
\end{aligned}$$

Hence
$$\begin{aligned}
r_{odq} &= [C]\, r_{abc}\, [C_t] \\
L_{odq} &= [C]\, L\,[C_t] \\
\omega G &= [C]\, p\{L\,[C_t]\} = [C]\left\{\frac{d}{d\theta}\, L\,[C_t]\right\}\frac{d\theta}{dt}\,.
\end{aligned}$$

The 'L' terms associated with di/dt provide the components of transformer e.m.f. for a given coil, whilst those associated with $d\theta/dt$ provide the components of rotational voltage. The complexity of the matrix depends on the number of coils on the real machine, (e.g. damper windings are often considered on both the d and q-axes). The flux linkages for a phase coil, due to i_a, i_b, i_c, determine the terms in the matrix:

$$[L_{abc}] = \begin{vmatrix} L_a & L_{ab} & L_{ac} \\ L_{ba} & L_b & L_{bc} \\ L_{ca} & L_{cb} & L_c \end{vmatrix}.$$

Then

$$[L_{odq}] = [C]\, L_{abc}\, [C_t]$$

$$= \begin{vmatrix} L_l & 0 & 0 \\ 0 & L_l + \frac{3}{2}(L_1 + L_2) & 0 \\ 0 & 0 & L_l + \frac{3}{2}(L_1 - L_2) \end{vmatrix}$$

$$= \begin{vmatrix} L_o & 0 & 0 \\ 0 & L_d & 0 \\ 0 & 0 & L_q \end{vmatrix}.$$

Thus, for a balanced system, but including the field coil:

$$[L_{fdq}] = \begin{vmatrix} L_f & M_d & 0 \\ M_d & L_d & 0 \\ 0 & 0 & L_q \end{vmatrix}$$

where $M_d = \sqrt{(3/2)}\, M$

Rotational voltage coefficients occur for the d and q coils only and thus we obtain the G-matrix, (see also Refs. 3, 4):

$$[G_{fdq}] = \left| [C] \frac{\mathrm{d}}{\mathrm{d}\theta} \{L[C_t]\} \right|$$

$$= \begin{vmatrix} 0 & 0 & 0 \\ 0 & 0 & -L_q \\ M_d & L_d & 0 \end{vmatrix}.$$

The voltage equations for the d–q model are thus:

$$\begin{vmatrix} e_f \\ e_d \\ e_q \end{vmatrix} = \begin{vmatrix} r_f + L_f p & M_d p & 0 \\ M_d p & r + L_d p & -\omega L_q \\ \omega M_d & \omega L_d & r + L_q p \end{vmatrix} \times \begin{vmatrix} i_f \\ i_d \\ i_q \end{vmatrix}$$

Steady-state operation

Under steady-state the p-terms vanish and:

$$e_d = i_d r - \omega L_q i_q$$
$$e_q = i_q r + \omega L_d i_d + \omega M_d i_f$$
$$e_f = i_f r_f .$$

Note that ωL_d and ωL_q are reactances and are defined as:

$\omega L_d = X_d =$ direct-axis synchronous reactance

$\omega L_q = X_q =$ quadrature-axis synchronous reactance.

The synchronous machine on symmetrical short-circuit (Ref. 6)

The voltage equations for a synchronous machine, in terms of d, q and f coils are given in the matrix above. Under conditions of short circuit we have:

$$e_q = e_d = 0,$$

and

$$e_f = i_f(r_f + L_f p) + M_d p i_d \quad (1)$$
$$0 = M_d p i_f + i_d(r + L_d p) - \omega L_q i_q \quad (2)$$
$$0 = \omega M_d i_f + \omega L_d i_d + i_q(r + L_q p) \quad (3)$$

Consider the first two equations;
If at $t = 0$, $i_q = 0$, $i_d = 0$ and $e_f = V_f$, (see oscillograms Fig. 6.30) then:

$$V_f - i_f r_f = L_f p i_f + M_d p i_d$$
$$0 = M_d p i_f + L_d p i_d .$$

Now at $t = 0$, $V_f - i_f r_f = 0$ and we can rearrange the equations:

$$L_f p i_f = -M_d p i_d$$
$$M_d p i_f = -L_d p i_d .$$

Cross multiplying and collecting terms:

$$(L_f L_d - M_d^2)(pi_f)(pi_d) = 0.$$

Hence $pi_f = pi_d = 0$, providing $(L_f L_d - M_d^2) \neq 0$, which is true for a practical machine since unity coupling is not achieved. Further, from equation (3), at $t = 0$, (and $i_q = i_d = 0$):

$$\omega M_d i_f = L_q p i_q$$

or

$$p i_q = \omega M_d i_f / L_q = E/L_q$$

where E = peak of induced a.c. at switching.

From the differential of equations (1) and (2) at $t = 0$ we also have $p^2 i_d = 0$ and $p^2 i_q = 0$. These conclusions are important later in determining the constants of the final exponential form of solution for the s.c. current.

Now from equation (1):

$$i_f = \left[\frac{e_f}{r_f + L_f p} - \frac{M_d p i_d}{r_f + L_f p}\right]. \tag{4}$$

Consider the first term of equation (4);

If we assume r_f negligible:

$$\frac{e_f}{r_f + L_f p} \approx \frac{e_f}{L_f p}.$$

But with $r_f = 0$, the final steady-state field current would be infinite! However, we maintain that the total field flux-linkages must remain constant. So we imply, (with r_f neglected):

$$i_f = \frac{e_f}{L_f p} - \frac{M_d p i_d}{L_f p}$$

or

$$\frac{e_f}{p} = L_f i_{f(t)} + M_d i_{d(t)} = L_f i_{f(t=0)}.$$

Using this approximation in equation (4), (restoring r_f):

$$i_f = (i_{fo} - i_{ft}) = \frac{M_d p i_d}{(r_f + L_f p)}.$$

Substituting in equations (2) and (3) and rearranging;

$$O = \left\{r + \left[L_d + \frac{(M_d^2)p}{r_f + L_f p}\right]p\right\} i_d - \omega L_q i_q \tag{5}$$

$$E = \left[L_d - \frac{(M_d^2)\,p}{(r_f + L_f\,p)}\right]\omega i_d + (r + L_q p)\,i_q . \qquad (6)$$

Define the quantity:

$$\left[L_d - \frac{(M_d^2)\,p}{r_f + L_f\,p}\right] = L'_d\,(p)$$

say $= L$ (for the moment).

$$O = (r + Lp)\,i_d - \omega L_q i_q$$
$$E = \omega L i_d + (r + L_q p)\,i_q .$$

Solve for i_d and i_q using determinants, or by using convenient coefficients, A, B, C, D:

$$O = Ai_d - Bi_q$$
$$E = Ci_d + Di_q$$

$$i_d = \frac{BE}{AD + BC} \text{ and } i_q = \frac{AE}{AD + BC} .$$

Now $(AD + BC) = (r + Lp)(r + L_q p) + \omega^2 LL_q$
$= r^2 + LL_q p^2 + r(L_q p + Lp) + \omega^2 LL_q .$

The term r^2 can be neglected since r is small. $\omega^2 LL_q$ can be written $X_q X'_d\,(p)$ since $L = L'_d\,(p)$ and is a function of p. Hence:

$$(AD + BC) = X_q X'_d(p)\left[\frac{p^2}{\omega^2} + \frac{r}{\omega}\left(\frac{1}{X'_d(p)} + \frac{1}{X_q}\right)p + 1\right].$$

If r_f is assumed negligible, $X'_d\,(p)$ is no longer a function of p and we can substitute:

$$X'_d = \omega\left(L_d - \frac{M_d^2}{L_f}\right) = d\text{-axis transient reactance}$$

$$X_d = \omega L_d = d\text{-axis synchronous reactance}$$

$$X_q = \omega L_q = q\text{-axis synchronous reactance.}$$

Now $T_f = L_f / r_f$, sometimes called the O/C d-axis transient time constant.

Let $\dfrac{X'_d}{X_d} T_f = T'_d$ and $\zeta = \dfrac{r}{2}\left(\dfrac{1}{X'_d} + \dfrac{1}{X_q}\right).$

Substituting values for A, B, C, D etc., we get:

$$(T'_d p + 1)\left(\frac{p^2}{\omega^2} + \frac{2\zeta p}{\omega} + 1\right) i_d = -(T_f p - 1)\frac{E}{X_d}$$

$$(T'_d p + 1)\left(\frac{p^2}{\omega^2} + \frac{2\zeta p}{\omega} + 1\right) i_q = 0.$$

Using the approximations of r and r_f being small, giving $\omega T'_d \gg 1$, $\omega T_f \gg 1$, and $\zeta \ll 1$, and the conditions obtained at $t = 0$, the solutions are obtained:

$$i_d = -\frac{E}{X_d} + \frac{E}{X'_d}\epsilon^{-\zeta\omega t}\cos\omega t - E\left(\frac{1}{X'_d} - \frac{1}{X_d}\right)\epsilon^{-t/T'_d}$$

$$i_q = \frac{E}{X_q}\epsilon^{-\zeta\omega t}\sin\omega t.$$

T'_d is called the d-axis transient S/C time constant.

Transformations to abc axes occurs using, (e.g.):

$$i_a = i_d\cos(\omega t + \delta) + i_q\sin(\omega t + \delta)$$

where δ corresponds to the instant of S/C. Then on S/C:

$$i_a = -\frac{E}{X_d}\cos(\omega t + \delta) + \frac{E}{2}\left(\frac{1}{X'_d} + \frac{1}{X_q}\right)\epsilon^{-\zeta\omega t}\cos\delta$$

$$+ \frac{E}{2}\left(\frac{1}{X'_d} - \frac{1}{X_q}\right)\epsilon^{-\zeta\omega t}\cos(2\omega t + \delta)$$

$$- E\left(\frac{1}{X'_d} - \frac{1}{X_d}\right)\epsilon^{-t/T'_d}\cos(\omega t + \delta)$$

= steady-state s/c current + transient d.c. part

+ second harmonic transient, (usually small)

+ damped sinusoid.

Expressions for the b-phase and c-phase currents are obtained by substituting:

$$\left(\delta - \frac{2\pi}{3}\right) \text{ and } \left(\delta + \frac{2\pi}{3}\right)$$

for δ in the expression for i_a. An approximate physical interpretation is given on p. 192.

REFERENCES

1. ADKINS, B. (1962). *The General Theory of Electrical Machines*, Chapman and Hall, London.
2. FITZGERALD, A.E. and KINGSLEY, C. (1961). *Electric Machinery*, 2nd. ed., McGraw-Hill, New York.
3. MESSERLE, H.K. (1965). *Dynamic Circuit Theory*, Pergamon Press, Oxford.
4. O'KELLY, D. and SIMMONS, S. (1968). *Introduction to Generalised Electrical Machine Theory*, McGraw-Hill, New York.
5. HANCOCK, N.N. (1964). *Matrix Analysis of Electrical Machinery*, Pergamon Press, Oxford.
6. MEISEL, J. (1966). *Principles of Electromechanical Energy Conversion*, McGraw-Hill, New York.

Appendix: The transformer

The ideal transformer

Consider two coils mutually coupled ideally. For a sinusoidal alternating mutual flux, $\phi = \Phi_m \sin \omega t$.

$$\text{Induced e.m.f., } e = -N\frac{\mathrm{d}\phi}{\mathrm{d}t} = -\omega\Phi_m N \cos \omega t.$$

This is a maximum when $\cos \omega t = 1$.

$$\therefore \quad \text{Induced } E_m = -\omega\Phi_m N = -2\pi f \Phi_m N.$$

$$\text{Then, r.m.s. value, } E = \frac{1}{\sqrt{2}} E_m = -4{\cdot}44\,\Phi_m fN \text{ volts.}$$

Thus for coil 1, termed the Primary:

$$\text{Induced e.m.f., } E_1' = -4{\cdot}44\,\Phi_m fN_1$$

$$\text{Secondary induced e.m.f., } E_2 = -4{\cdot}44\,\Phi_m fN_2 \,.$$

The corresponding primary applied voltage:

$$E_1 = +4{\cdot}44\,\Phi_m fN_1 \text{ volts.}$$

Thus N_1/N_2 = turns ratio, k, = E_1/E_2.
Note that:

$$\Phi_m = \frac{E_1}{4{\cdot}44\,fN_1} = \text{constant for constant } E_1.$$

Now, if a load is connected to coil 2, a current will be drawn from the supply to the primary, such that:

$$I_1 N_1 = I_2 N_2$$

because Φ_m is constant. The fact may also be explained by the Principle of Constant Flux-linkages, viz. if no balancing m.m.f. occurred in the primary, an additional e.m.f. would be induced in contradiction of Kirchhoff's voltage law.

Hence we have: $\dfrac{N_1}{N_2} = \dfrac{E_1}{E_2} = \dfrac{I_2}{I_1} = k$, the turns ratio.

A phasor diagram, and a time-scale diagram can be drawn as in Fig. A.1.

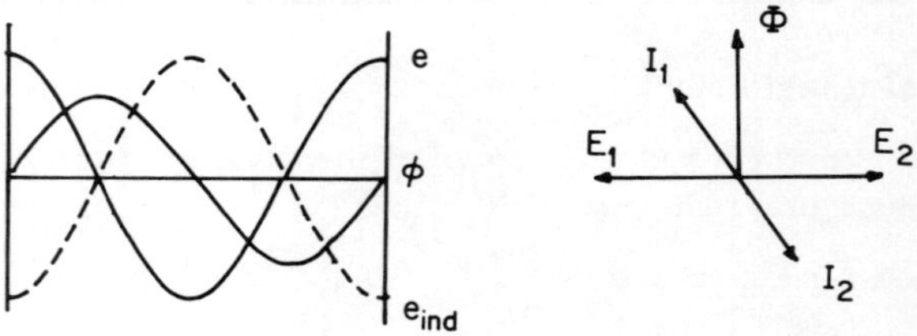

Fig. A.1.

The practical transformer

In practice (a) coupling is assisted by providing a common iron core linking the two coils. However, ideal coupling is not achieved, and some of the primary flux follows a leakage path. Similarly, with a secondary load, some of the secondary m.m.f. is lost due to leakage flux paths about the secondary winding, Fig. A.2. (b) the windings have ohmic resistance. (c) on no-load condition, energy is taken from the supply to provide the 'core losses', arising from hysteresis and eddy current effects in the iron core. The design aims to minimise this core loss by (*i*) using a laminated core construction, (*ii*) employing a grade of iron alloy which has a low hysteresis loss, but with a high μ_r to minimise the magnetising current necessary for a given flux.

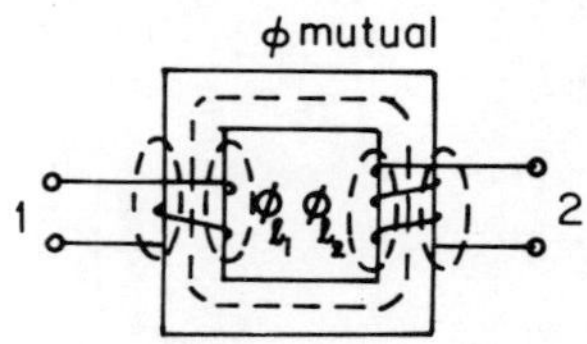

Fig. A.2.

These departures from the ideal transformer can be allowed for by lumped components, and an equivalent circuit for a practical transformer can be drawn as in Fig. A.3.

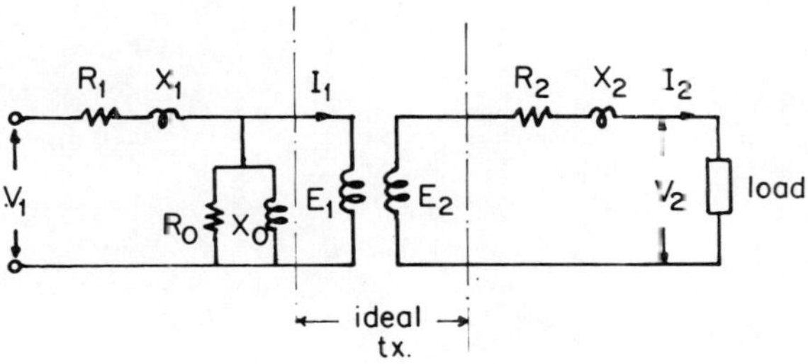

Fig. A.3.

The no-load condition

On no load at normal voltage V_1 applied, a transformer takes a current I_0 lagging V_1 by θ_0. Then power input on no load, $W_0 = V_1 I_0 \cos\theta_0$.

$I_0 \cos\theta_0 = I_e$ is termed the energy component of the no-load current,

$I_0 \sin\theta_0 = I_m$ is termed the magnetising component, and is in phase with Φ_m.

Thus I_0 is interpreted as two components and the no-load condition can be represented by an equivalent circuit of R_0 in parallel with X_0 where $R_0 = V_1/I_e$ and $X_0 = V_1/I_m$.

Since R_1 and X_1 are found to be $\ll R_0$ and X_0, the no-load equivalent circuit may be placed at the terminals, Fig. A.4.

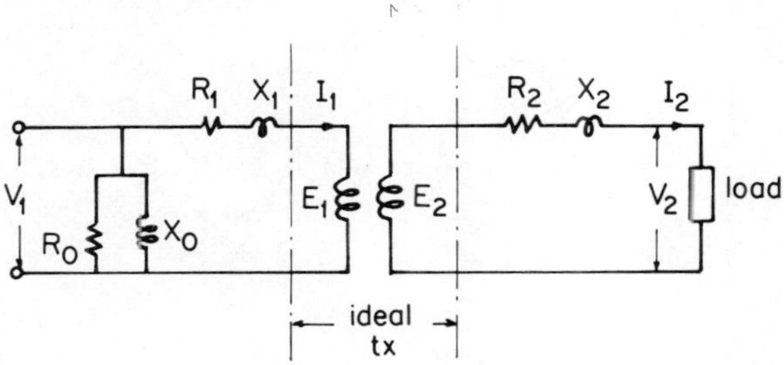

Fig. A.4.

The core losses:

It has been shown in Chapter 3 that the hysteresis loss $= k_h B^x f$, and that due to eddy currents $= k_e B^2 f^2$.

$$\text{Thus core loss, } W_0 = W_h + W_e$$
$$= k_h B^x f + k_e B^2 f^2 \text{ Watts.}$$

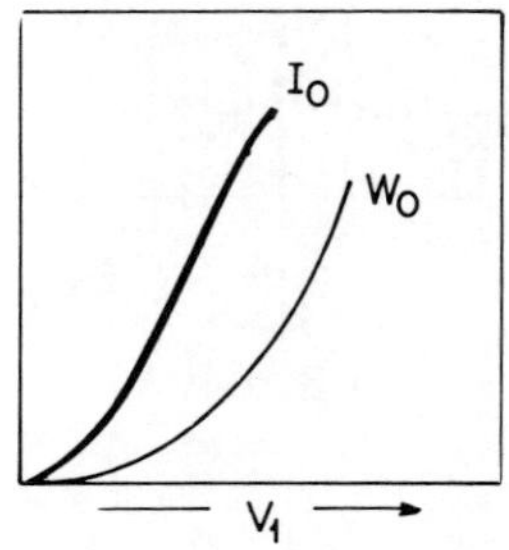

Fig. A.5.

(*i*) A test carried out on a transformer on no load, but for various applied voltages provides a variation in W_0 which will approximate to a square law, Fig. A.5. This is because for transformer sheet steel, commonly used for cores, $x = 1{\cdot}6 - 1{\cdot}8$ so that W_0 approx. $\propto B^2 \propto V^2$. In the same test I_0 will exhibit the non-linearities characteristic of a magnetisation curve. The result is that I_0 has a harmonic content, mainly 3rd harmonic, Fig. A.6. However it is only a small percentage of load current and may generally be neglected in calculations unless harmonics are an important issue.

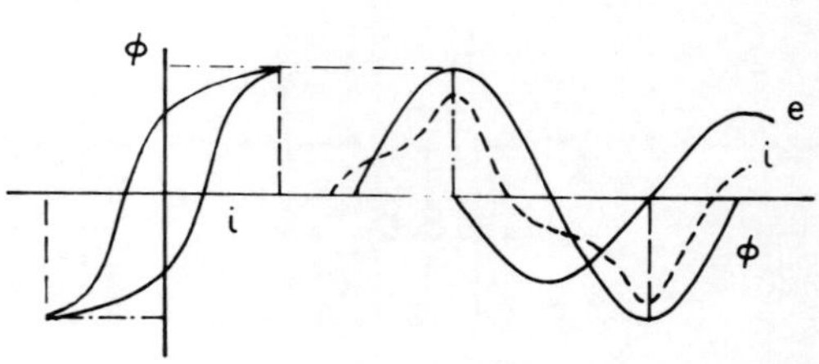

Fig. A.6.

(*ii*) Separation of losses. If flux is maintained constant, then:

$$W_0 = k_1 f + k_2 f^2.$$

Hence a no-load test can be made, observing W_0 for different values of supply frequency, but maintaining constant flux by keeping V/f constant. Then by plotting W/f against f a straight line relation is obtained from which k_1 and k_2 can be determined, Fig. A.7. Hence the hysteresis and eddy-current components of W_0 can be separated.

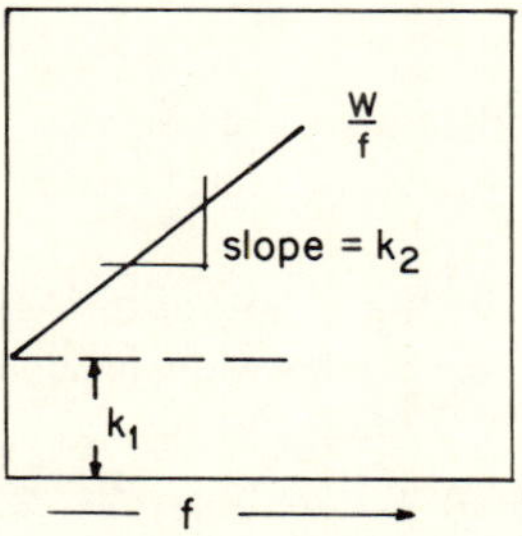

Fig. A.7.

Referred values of resistance and reactance

It is frequently convenient to view the transformer as a 'black box', having a total effective impedance. We may wish to consider all the resistances and reactances referred to the supply side, (primary), or to the load side, (secondary).

Now 1 volt in the secondary = k volts in the primary

$I_2 R_2$ volts in the secondary = $k(I_2 R_2)$ volts in the primary.

But the current in the primary = I_1 and $I_2/I_1 = k$.

Hence we can say, $I_2 R_2$ volts, (secondary) = $I_1 k^2 R_2$ volts (primary).

Similarly, $I_2 X_2$ volts (secondary) = $I_1 k^2 X_2$ volts (primary).

Thus, total R referred to primary = $R' = R_1 + k^2 R_2$

total X referred to primary = $X' = X_1 + k^2 X_2$.

An equivalent circuit, (referred to primary, for example) can be drawn, Fig. A.8.

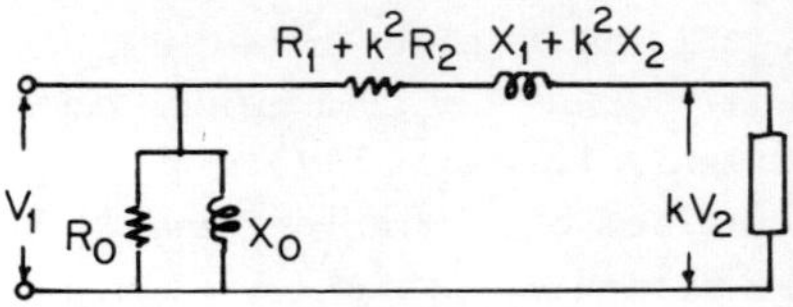

Fig. A.8.

The short-circuit test

The above 'total' equivalent circuit can be evaluated from results obtained from no-load, (already considered), and short-circuit tests.

In the s/c test care is taken to apply a much reduced voltage to the primary since a s/c connection is made at the secondary. Readings are taken of V_{sc}, I_{sc} and W_{sc}.

Now, with reduced applied voltage, W_0 is negligible, (Fig. A.5). Hence:

$$W_{sc} = I_{sc}^2 R'.$$

But $V_{sc}/I_{sc} = Z'$. Hence, $X' = (Z'^2 - R'^2)^{\frac{1}{2}}$.

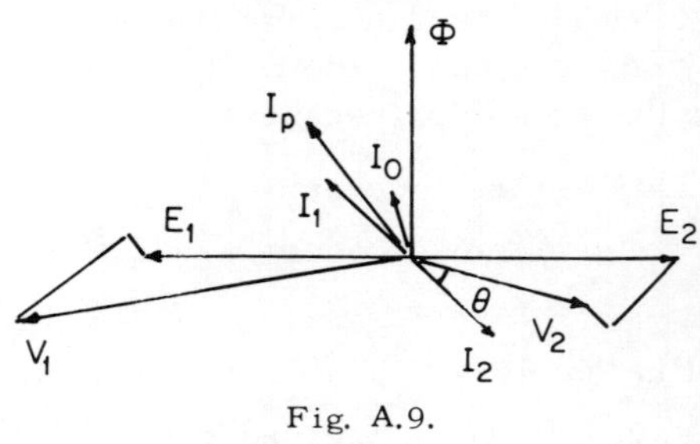

Fig. A.9.

The on-load phasor diagram

Referring to the equivalent circuit diagram, Fig. A.3 and also to Fig. A.9, we may start with the imposed secondary load conditions, say V_2, I_2 and θ_2. Adding $I_2 R_2$ and $I_2 X_2$ to V_2, in correct phase relations determines E_2 the ideal induced secondary voltage. E_1 is anti-phase to E_2 and Φ_m is in quadrature with both. I_1 will be anti-phase to I_2.

Introduce I_0 leading upon Φ_m. Then I_p, the total primary current is the phasor sum of I_1 and I_0. Allow for $I_p R_1$ and $I_p X_1$, whereby V_1 is determined in magnitude and relative phase.

Note (i) a 1/1 ratio is implied for graphical convenience, (ii) the addition of phasors is usually an easier task than subtraction. But having established the phasor diagram as above, it may of course be explained by starting at V_1.

It will be seen that it is seldom necessary to draw the diagram to scale, but an understanding of it is necessary.

Voltage regulation

This is defined as the change in terminal voltage due to loading. Expressed in percentage or per unit terms it is not then important to which side the calculation is referred.

In phasor notation:

$$\mathbf{V} = \mathbf{E} - \mathbf{I}(\mathbf{R}' + j\mathbf{X}')$$

$$\therefore \quad \text{Regulation} = \frac{\mathbf{IZ}}{\mathbf{V}} \text{ per unit.}$$

An interesting and popular alternative method is based on the geometry of the phasor diagram, as follows:

$$\begin{aligned} \text{Regulation} &= \mathbf{E} - \mathbf{V} \\ &\simeq |E| - |V| \\ &\simeq (IR' \cos\theta \pm IX' \sin\theta) \text{ volts.} \end{aligned}$$

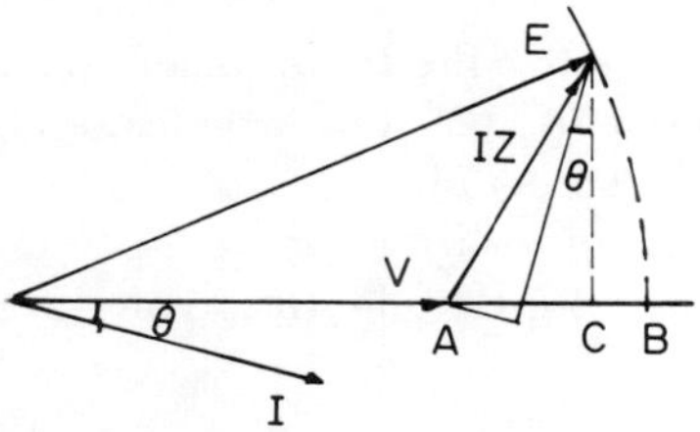

Fig. A.10.

The two graphical approximations are possible because in a transformer the volt-drop triangle is very small compared with V, (< 5%). In Fig. A.10 its size is exaggerated to emphasize the graphical approximations involved. Thus:

$$\mathbf{IZ} \simeq AB \simeq (IR' \cos\theta + IX \sin\theta).$$

Note that if θ is a leading angle, $(IX' \sin\theta)$ overlaps $(IR' \cos\theta)$ and the $-ve$ alternative sign appears before the $(IX' \sin\theta)$ term as above.

Efficiency

The efficiency, $$\eta = \frac{\text{Output}}{\text{Output} + \text{losses}}$$

$$= \frac{VI\cos\theta}{VI\cos\theta + W_0 + I^2R'}\,.$$

Since V is substantially constant, load current and load are proportional. Hence for maximum efficiency, $d\eta/dI = 0$, and thus:

$$\frac{d\eta}{dI} = \frac{(VI\cos\theta + I^2R' + W_0)\,V\cos\theta - (VI\cos\theta)(V\cos\theta + 2IR')}{(\text{denominator})^2}$$

$= 0$ for maximum

$\therefore$ $\underline{I^2R' = W_0}$ is the condition for maximum efficiency.

Note: copper loss $\propto I^2$

core loss is constant for constant V

output $\propto$ power factor.

Transformers in parallel

It will be seen that ideally the transformers should have identical transformation ratios and identical impedance characteristics, if the load is to be shared equally.

Case 1. Identical transformation ratios:

Parallel operation imposes identical voltage regulation, Fig. A.11(a), hence:

$$v = \mathbf{I}_A\mathbf{Z}_A = \mathbf{I}_B\mathbf{Z}_B = \mathbf{I}\left(\frac{\mathbf{Z}_A\mathbf{Z}_B}{\mathbf{Z}_A + \mathbf{Z}_B}\right)$$

$$\therefore \quad \mathbf{I}_A = \mathbf{I}\left(\frac{\mathbf{Z}_B}{\mathbf{Z}_A + \mathbf{Z}_B}\right)$$

$$\mathbf{I}_B = \mathbf{I}\left(\frac{\mathbf{Z}_A}{\mathbf{Z}_A + \mathbf{Z}_B}\right).$$

Note that the load current is shared in the inverse ratio of the impedances.

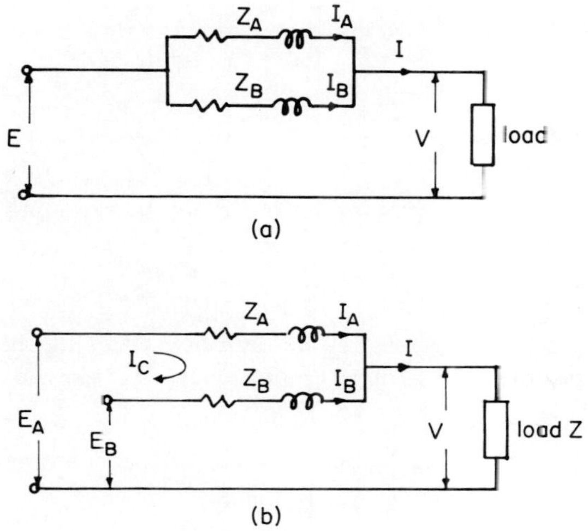

Fig. A.11.

Since the transformers share a common terminal voltage, the result may be expressed in terms of shared kVA, (symbol Q),

$$\mathbf{Q}_A = \mathbf{Q}\left(\frac{\mathbf{Z}_B}{\mathbf{Z}_A + \mathbf{Z}_B}\right)$$

$$\mathbf{Q}_B = \mathbf{Q}\left(\frac{\mathbf{Z}_A}{\mathbf{Z}_A + \mathbf{Z}_B}\right)$$

and $\mathbf{Q} = (kW \pm j\, kVAR)$ in symbolic notation.

Case 2.

This is the case where the transformation ratio may be slightly different for the two transformers. The equivalent circuit, referred to the common-load side is shown in Fig. A.11(b):

$$\mathbf{I} = \mathbf{I}_A + \mathbf{I}_B .$$

Then
$$\mathbf{E}_A - \mathbf{I}_A\mathbf{Z}_A = (\mathbf{I}_A + \mathbf{I}_B)\,\mathbf{Z}$$
$$\mathbf{E}_B - \mathbf{I}_B\mathbf{Z}_B = (\mathbf{I}_A + \mathbf{I}_B)\,\mathbf{Z}.$$

Rearranging:

$$\mathbf{I}_A(\mathbf{Z} + \mathbf{Z}_A) + \mathbf{I}_B\mathbf{Z} = \mathbf{E}_A$$
$$\mathbf{I}_A\mathbf{Z} + \mathbf{I}_B(\mathbf{Z} + \mathbf{Z}_B) = \mathbf{E}_B$$

$$\therefore \quad \mathbf{I}_A = \frac{\mathbf{E}_A(\mathbf{Z} + \mathbf{Z}_B) - \mathbf{E}_B\mathbf{Z}}{(\mathbf{Z} + \mathbf{Z}_A)(\mathbf{Z} + \mathbf{Z}_B) - \mathbf{Z}^2}$$

$$= \frac{\mathbf{Z}(\mathbf{E}_A - \mathbf{E}_B) + \mathbf{Z}_B\mathbf{E}_A}{\mathbf{Z}_A\mathbf{Z}_B + \mathbf{Z}\mathbf{Z}_A + \mathbf{Z}\mathbf{Z}_B}.$$

A similar expression is obtained for $\mathbf{I}_B$ by transposing A and B.

The unbalance, or circulating current:

$$\mathbf{I}_c = (\mathbf{E}_A - \mathbf{E}_B)/(\mathbf{Z}_A + \mathbf{Z}_B).$$

An alternative solution may be obtained by using the superposition theorem, and first assuming equal transformation ratios, as in Case 1, giving, $\mathbf{I}'_A$ and $\mathbf{I}'_B$. Then:

$$\mathbf{I}_A = \mathbf{I}'_A + \mathbf{I}_c$$
$$\mathbf{I}_B = \mathbf{I}'_B - \mathbf{I}_c.$$

It is worth remembering that these solutions, in principle, may be applied to any two devices operating in parallel, where R and X may be assumed constants, e.g. feeders in parallel, and in simplified examples of synchronous machines in parallel.

EXAMPLES

EXAMPLE A.1.

Explain, by reference to fundamental theory, the factors which determine the choice of a suitable material for the core of a power transformer.

A single-phase transformer has 480 primary turns and 90 secondary turns. The mean length of the flux path in the core is 180 cm and the joints are equivalent to an air gap of 0·1 mm. If the maximum value of the core flux density is 1·1 Wb/m^2 with 2200 V at 50 c/s applied to the primary, calculate the primary current and power factor on no load.

Assume 400 ampere-turns/metre at 1·1 Wb/m^2 and the corresponding iron loss 1·76 W/kg at 50 c/s; density of the iron 7·8 gm/cm^3.

Solution:

First part. Reference to hysteresis and eddy current loss considerations, (see p. 52) demonstrates the need to use a low loss iron alloy, preferably laminated, in the core construction. Also the flux density at which saturation occurs in a given magnetic material determines either, the working flux density or the desirable cross section to be used in the core.

Second part. Amp-turns for iron core $= 400 \times 1{\cdot}8 = 720$ AT

Amp-turns for airgap $= Bl/\mu_0$

$$= \frac{1{\cdot}1 \times 10^7 \times 0{\cdot}0001}{4\pi} = 87{\cdot}5 \text{ AT}$$

$\therefore$ Total AT $= 720 + 87{\cdot}5 = 807{\cdot}5$ AT.

Hence, magnetizing current, I_m (max) $= 807{\cdot}5/480$

$= 1{\cdot}682$ A.

$\therefore \quad I_m$ (r.m.s.) $= 1{\cdot}19$A.

Now $\quad E_1 = 4{\cdot}44\, \Phi_{max}\, f\, T_1$

$$\therefore \quad \Phi_{max} = \frac{2200}{4{\cdot}44 \times 50 \times 480} = 0{\cdot}0206 \text{ Wb}$$

$$\therefore \quad \text{c.s.a. of core} = \Phi_{max}/B_{max} = \frac{0{\cdot}0206}{1{\cdot}1} = 0{\cdot}0187 \text{ m}^2$$

$= 187$ cm^2.

Volume of iron $= 187 \times 180 = 33\,700$ cm^3

Weight of iron $= 33\,700 \times 7{\cdot}8 = 262{\cdot}5$ kg

$\therefore$ Iron loss $= 262{\cdot}5 \times 1{\cdot}76 = 464$ W.

$$\therefore \quad \text{Energy component of current, } I_e = \frac{464}{2200} = 0{\cdot}21\text{A}.$$

Hence, no load current, $I_0 = (I_e^2 + I_m^2)^{\frac{1}{2}} = 1{\cdot}21$A

$\cos\phi_0 = I_e/I_0 = 0{\cdot}174$ lag.

EXAMPLE A.2.

Two 200 kVA single-phase transformers of identical transformation ratios operate in parallel to share a load of 320 kW, 0·8 p.f. lag. The impedances of the transformers are

$Z_A = (0{\cdot}5 + j8)$ and $Z_B = (0{\cdot}75 + j4)$ respectively. Calculate the load carried by each transformer.

Solution:

$$Z_A = (0{\cdot}5 + j8) = (64{\cdot}25)^{\frac{1}{2}} \underline{/\tan^{-1} 8/0{\cdot}5} = 8{\cdot}1 \underline{/86{\cdot}4^\circ}$$

$$Z_B = (0{\cdot}75 + j4) = (16{\cdot}56)^{\frac{1}{2}} \underline{/\tan^{-1} 4/0{\cdot}75} = 4{\cdot}06 \underline{/79{\cdot}4^\circ}$$

$$Z_A + Z_B = (1{\cdot}25 + j12) = (146{\cdot}56)^{\frac{1}{2}} \underline{/\tan^{-1} 12/1{\cdot}25} = 12{\cdot}05 \underline{/84{\cdot}1^\circ}.$$

$$\text{Total load} = Q = \frac{320}{0{\cdot}8} \underline{/-36{\cdot}9^\circ} = 400 \underline{/-36{\cdot}9^\circ}$$

$$\therefore \quad Q_A = Q \frac{Z_B}{Z_A + Z_B} = \frac{400 \times 4{\cdot}06}{12{\cdot}05} \underline{/-36{\cdot}9 + 79{\cdot}4 - 84{\cdot}1^\circ}$$

$$= 134{\cdot}4 \underline{/-41{\cdot}6^\circ} \text{kVA}$$

$$= 100 \text{ kW @ p.f. } 0{\cdot}748 \text{ lagging.}$$

$$Q_B = Q \frac{Z_A}{Z_A + Z_B} = \frac{400 \times 8{\cdot}1}{12{\cdot}05} \underline{/-36{\cdot}9 + 86{\cdot}4 - 84{\cdot}1^\circ}$$

$$= 268{\cdot}4 \underline{/-36{\cdot}6^\circ} \text{kVA}$$

$$= 220 \text{ kW @ p.f. } 0{\cdot}82 \text{ lagging.}$$

N.B. As a check, note that the kW may be added arithmetically, but the kVA add as phasors.

Index